POLLUTION SCIENCE, TECHNOLOGY AND ABATEMENT

THE ROLE OF SORBENTS IN SAMPLING AND ANALYSIS OF EMERGING POLLUTANTS IN INDOOR ENVIRONMENTS

POLLUTION SCIENCE, TECHNOLOGY AND ABATEMENT

Additional books in this series can be found on Nova's website under the Series tab.

Additional E-books in this series can be found on Nova's website under the E-books tab.

ENVIRONMENTAL SCIENCE, ENGINEERING AND TECHNOLOGY

Additional books in this series can be found on Nova's website under the Series tab.

Additional E-books in this series can be found on Nova's website under the E-books tab.

POLLUTION SCIENCE, TECHNOLOGY AND ABATEMENT

THE ROLE OF SORBENTS IN SAMPLING AND ANALYSIS OF EMERGING POLLUTANTS IN INDOOR ENVIRONMENTS

CARMEN GARCIA-JARES
JORGE REGUEIRO
MARÍA LLOMPART
AND
RUTH BARRO

Nova Science Publishers, Inc.
New York

Copyright © 2010 by Nova Science Publishers, Inc.

All rights reserved. No part of this book may be reproduced, stored in a retrieval system or transmitted in any form or by any means: electronic, electrostatic, magnetic, tape, mechanical photocopying, recording or otherwise without the written permission of the Publisher.

For permission to use material from this book please contact us:
Telephone 631-231-7269; Fax 631-231-8175
Web Site: http://www.novapublishers.com

NOTICE TO THE READER

The Publisher has taken reasonable care in the preparation of this book, but makes no expressed or implied warranty of any kind and assumes no responsibility for any errors or omissions. No liability is assumed for incidental or consequential damages in connection with or arising out of information contained in this book. The Publisher shall not be liable for any special, consequential, or exemplary damages resulting, in whole or in part, from the readers' use of, or reliance upon, this material.

Independent verification should be sought for any data, advice or recommendations contained in this book. In addition, no responsibility is assumed by the publisher for any injury and/or damage to persons or property arising from any methods, products, instructions, ideas or otherwise contained in this publication.

This publication is designed to provide accurate and authoritative information with regard to the subject matter covered herein. It is sold with the clear understanding that the Publisher is not engaged in rendering legal or any other professional services. If legal or any other expert assistance is required, the services of a competent person should be sought. FROM A DECLARATION OF PARTICIPANTS JOINTLY ADOPTED BY A COMMITTEE OF THE AMERICAN BAR ASSOCIATION AND A COMMITTEE OF PUBLISHERS.

LIBRARY OF CONGRESS CATALOGING-IN-PUBLICATION DATA

Available upon Request
ISBN: 978-1-61761-004-2

Published by Nova Science Publishers, Inc. New York

CONTENTS

Preface vii
Chapter 1 Introduction 1
Chapter 2 Sampling Techniques and Sorbent Materials 5
Chapter 3 Desorption and Sample Treatment 19
Chapter 4 Determination Techniques 21
Chapter 5 Method Quality Evaluation 25
Chapter 6 Selected Applications 33
Chapter 7 Conclusion 79
References 81
General Abbreviations 97
Index 103

PREFACE

In the last few years the concern about pollutants present in indoor environments has greatly increased since in developed countries people spend long time indoors. The markedly growing consumption of cosmetics, air fresheners, house-cleaners, biocides, as well as the increase in the use of new building materials in carpeting, paints, furnishings, etc, may turn our homes, schools, offices and workplaces into harmful microenvironments. Inadequate ventilation coupled with the slow indoor degradation processes may increase indoor pollution levels. High temperature and humidity levels can also increase concentrations of some pollutants. Hence, inhalation of indoor air is potentially the most important exposure pathway to many pollutants.

The chemicals that are extensively found in indoor environments include compounds that are suspected to behave as priority organic pollutants and endocrine disrupting compounds such as phthalate esters, polybrominated and phosphate flame retardants, fragrances, pesticides, biocides, and other organic compounds such as organotin and perfluorinated alkyl compounds.

This book reviews and discuses recent literature related to methodological developments for the analysis of pollutants in indoor air, focusing the attention on emerging contaminants and biocides that can be found both in the air gas phase as well as associated to the suspended particulate matter and settled dust. Available sorbents for sample collection, as well as analyte desorption techniques, clean-up procedures, determination techniques, and method performance evaluation will be summarized and discussed.

Chapter 1

INTRODUCTION

The concern about pollutants present in indoor environments has increased in the last few years. Chemicals that building occupants are exposed to today are substantially different from those that occupants experienced years ago [1]. The markedly growing consumption of chemical products as cosmetics, air fresheners, house-cleaners, biocides, appliances and electronic equipment, building materials e.g. carpeting, paints, furnishings, etc. may turn our homes, schools, offices and workplaces into harmful microenvironments. People in developed countries spend up to about 90% of their time indoors [2, 3], and hence the exposure to household pollutants has dangerously increased. Thus, and taking into account that each person inhales about 22 m^3 air per day [4], inhalation of indoor air has potentially become the most important exposure pathway to many pollutants and can be more relevant to human exposure assessment than ambient concentrations. Concerning infants and most vulnerable people, although outdoor air pollution first brought the issue of air pollution health effects to public attention, it is now indoor air pollution that likely has the greatest impact e.g. on children´s health [5]. As a first approach to evaluate the relationship between indoor air pollution and human exposure to pollutants, the AIRMEX project started in 2003. Personal exposures were conducted with employers and/or teachers working in different occupational environments such as public buildings, schools and kindergartens. Preliminary results indicate that personal exposure concentrations are higher than the indoor/outdoor concentrations [6].

Air is considered a very difficult environmental matrix to handle. It is a heterogeneous system of gases, aerosols and solid particles and its composition evolves, leading to a continuous movement, diffusion and reaction of the pollutants [7]. The concentration of an indoor pollutant depends not only on its indoor emission rate, but also on the rate at which it is being transported from outdoors to indoors, adsorbed by indoor surfaces, degraded by a slower indoor chemistry and removed by ventilation. Moreover, high temperature and humidity levels can increase concentrations of some pollutants. For example, it is well known that semivolatile organic compounds (SVOCs) are redistributed from their original sources to all indoor surfaces [8]. Vehicle garages can also be a pollution source as they often contain high concentrations of volatile organic compounds (VOCs) that may migrate into adjoining residences [9]. Surprisingly, some studies have identified indoor sources as the predominant factor influencing outdoor ambient air concentrations in densely populated areas [10].

Compounds that can be found indoors are flame retardants as polybrominated diphenyl ethers (PBDEs) and biphenyls, organophosphates, plasticizers such as phthalates or organotin compounds, fragrances, pesticides like organophosphate compounds, pyrethroids, among many others. Some of them are suspected to be POPs and/or EDCs, and can be found in the gaseous phase and/or associated to the suspended particulate matter and settled dust. A recent review summarizes available information on emission rates and indoor concentrations of various pollutants, such as VOCs, ozone, particulate matter and SVOCs (phthalate esters, brominated flame retardants –BFRs-, organophosphate flame retardants and polycyclic aromatic hydrocarbons -PAHs) that are related to electronic office equipment e.g. computers, printers, and photocopy machines [11]. An important finding is that personal exposures may be significantly larger than those estimated through average pollutant indoor concentrations, due to proximity of users to the sources over extended periods of time. At this time, exposure levels to these chemicals are largely un-documented as a result of a gap in the regulatory requirements, and data of major exposure sources and pathways is extremely limited yet.

Measurement of organic pollutants in air is often a hard task to overcome, in part because of the large number of diverse compounds of potential concern, the variety of available techniques for sampling and analysis, and the lack of standardized methods. Consequently, the growing demand on environmental monitoring by the society, and the appearance on stage of new chemicals have encouraged the development of new, more rapid and sensitive, easy of use, and less expensive methods for the analysis of emerging pollutants in indoor air.

International agencies have published analytical methods for air monitoring, which are available for all users on their respective webpages.

But, standardization is a long process, and for that reason, in many cases, the proposed methods are long and tedious conventional methods, not upgraded to the upcoming and novel techniques. On the other hand, they constitute a very useful and valuable resource for routine analysis or for laboratories that perform air analysis occasionally. U. S. Environmental Protection Agency (US EPA) has published a compendium of methods for the determination of toxic organic compounds in ambient air [12]. Table 1 summarizes the methods for the analysis of pollutants of concern in air, involving the utilization of sorbents. On the part of the National Institute for Occupational Safety and Health (NIOSH), individual analytical methods have been listed by chemical name or method number [13]. In addition, about 100 standards which address analysis of workplace air samples have been developed by American Society for Testing and Materials (ASTM), and are available after payment [14]. Regarding the European Union, the Technical Committee CEN/TC 264 "Air Quality" of the European Committee for Standardization (CEN) has also published several European Standards for air analysis [15].

In this book, the role of sorbent materials in the analytical developments and applications, regarding emerging contaminants in the indoor environment are reported. Available sorbents for sample collection, as well as analyte desorption techniques, clean-up procedures, determination systems, and method performance evaluation are summarized and discussed.

Table 1. Brief description of EPA methods for air analysis involving the use of sorbents (from ref. 12)

Method no.	Applicable compounds	Sampling device	Desorption technique	Chromatographic technique
TO-1	VOCs	Active sampling on a Tenax cartridge	Thermal desorption	GC/MS, GC/FID
TO-2	VOCs	Carbon molecular sieve	Thermal desorption	GC/MS, GC/FID
TO-3	VOCs	Cryogenic preconcentration in a stainless steel trap packed with silanized glass beads	Cryofocusing	GC/FID, GC/ECD
TO-4A	Pesticides, PCBs	Active sampling using a filter and a PUF trap	Solvent extraction with 10% diethyl ether/hexane	GC/FID-ECD, GC/MS
TO-7	Anilines (N-Nitrosodimethylamine)	Active sampling through a cartridge containing Thermosorb/N adsorbent	Solvent extraction with dichloromethane	GC/MS
TO-9A	Dioxins	Active sampling using a high-volume sampler through a glass fiber filter and a PUF adsorbent cartridge	Solvent extraction with toluene	HRGC/HRMS
TO-10A	Pesticides, PCBs	Active sampling using a low-volume sampler through a PUF plug	Solvent extraction with 5% diethyl ether/hexane	GC coupled to multi-detectors (ECD, PID, FID, etc.)
TO-11A	Aldehydes, ketones	Active sampling with a coated DNPH-cartridge	Solvent extraction with acetonitrile	HPLC/UV
TO-13A	PAHs	Active sampling using a high-volume sampler through a glass fiber filter and a PUF or XAD-2 cartridge	Solvent extraction using 10% diethyl ether	GC/MS
TO-17	VOCs	Active sampling through a multi-bed sorbent tube e.g. TenaxGR/Carbopack B, Carbopack B/Carbosieve SIII, Carbopack B/Carboxen 1000, Carbopack C/Carbopack B/Carbosieve SIII, Carbopack C/Carbopack B/Carboxen 1000, etc.	Thermal desorption	GC/MS, GC/FID, etc.

Chapter 2

SAMPLING TECHNIQUES AND SORBENT MATERIALS

A proper sampling device should fulfil the requirement of providing a representative air sample. Enrichment into solid sorbents is by far, the most widely used technique for sampling pollutants in air, although nowadays, other more recent techniques, such as solid-phase microextraction (SPME) or the use of membranes play an ever-increasing role.

Organic pollutants can be sampled in indoor air by whole, active, or passive sampling techniques. Major advantages and drawbacks of each sampling technique are summarized in table 2.

The simplest way to collect air samples is the whole air sampling, also called grab sampling, using polymer badges, glass containers or stainless-steel canisters. Total air sample is collected, avoiding breakthrough problems. Once the samples are collected, an aliquot can be introduced in a chromatographic system, either by direct injection, or using a preconcentration step, such as a cold trap or a cryofocusing device, which enhances the sensitivity of the method. Thus, some advantages derived from this procedure are that multiple aliquots of the sample can be re-injected into the system, and that very low blank levels are usually obtained. In addition, samples can be stored for several weeks without changes in their composition. However, these vessels may add a potential source of contamination, so they must be carefully pre-treated and pre-conditioned in order to avoid contamination or losses. The sample may contain a significant amount of water, which should be removed before analysis, resulting in evaporative losses of the less volatile compounds [16].

Table 2. Some advantages and disadvantages of air sampling methods

Type of sampling	Advantages	Disadvantages
Whole sampling	Simple Total sampling No breakthrough No degradation No moisture effect Low blank levels Long storage	Need a preconcentration step to achieve acceptable detection limits Contamination by the inner surfaces of the vessel Possible irreversible losses due to wall adsorption Possible losses while water removing before analysis Need pretreatment and preconditioning Severe clean-up procedures between samples Expensive to transport: heavy bottles Expensive devices: inner bottle surface and the clean-up step of shut-off valves
Passive sampling	Simple Cheap Long-term exposures Simultaneous deployment in several locations	Unstable flow-rates Influenced by meteorological conditions Long sampling times Difficult calibration
Active sampling	Short-term exposures Suitable for a wide volatility range of analytes Easy calibration Re-utilization of sorbents	Careful sorbent selection Measure the breakthrough volumes Need pumps Expensive Possible degradations Interferences with moisture

The use of passive air samplers for indoor and workplace air is not as common as active sampling. Passive air samplers are increasingly employed for monitoring POPs due to their ideal applicability for long-term on site monitoring providing TWA estimations. The currently available passive sampling devices are either based on permeation or diffusion. Because of their simplicity, low prize and their ease of use, diffusive sampling devices became very popular the last few years. Nevertheless, their low sampling rates require long sampling times, from days to even several weeks at low concentrations, but it is low cost facilitates simultaneous deployment in a large number of locations. The most relevant drawback for this type of sampling is that environmental conditions such as temperature, humidity, wind, and air velocity influence the quantification significantly. A further problem is the accumulation of artefacts in stored samples. Current calibration methods that exist for passive sampling, including equilibrium extraction, linear uptake, and kinetic calibration, are presented in the review written by Ouyang and Pawliszyn [17].

Passive samplers usually consist of the sampling medium (adsorbent layer), a diffusion path, and a diffusion barrier (permeable membrane) [18]. The receiving phase of a passive sampler can be solvent, polymer resin, chemical reagent or porous adsorbent [17]. Passive samplers include solvent-filled devices, semi-permeable membrane devices (SPMDs), membrane-enclosed sorptive coating sampler (MESCO), passive in situ concentration/extraction samplers (PISCES), sorbent-filled devices, polyurethane foam (PUF) disks and SPME devices [17]. Some of the most common passive sampling devices commercially available for air sampling are Radiello (patented by an Italian foundation), Analyst, ORSA-5 (Drägerwerk, Germany) and OVM (3 M, Germany) [19], all containing an adsorbing cartridge of activated charcoal. Empore disks (Waters), with an octadecyl (C_{18}) resin as the absorption material is another passive sampler which can be used for air analysis, and it is recommended for applications where high-pressure liquid chromatography (HPLC) is utilized for subsequent analysis.

SPMDs filled with triolein have become the most popular passive system for hydrophobic compounds. A problem is the time-consuming sample-treatment procedure by dialysis, necessary when SPMD is used [20]. SPMDs have been lately used e.g. for the determination of organophosphorus pesticides in the air of a research laboratory [21]. In this work, the analytes were quantitatively recovered by using a shorter extraction procedure using microwaves instead of the conventional process by dialysis.

As they have been increasingly in use for passive sampling, it is worth a mention the widely used PUF disks, applied e.g. for the collection of brominated flame retardants [22, 23], perfluorinated compounds (PFCs) [24], or polychlorinated byphenyls (PCBs) and organochlorine pesticides [25]. Last year, a comprehensive study was published on the subject [26]. PUF disks are generally mounted inside two stainless steel bowls to buffer the air flow to the disk and to shield it from precipitation and light [27]. Semivolatile compounds such as PCBs, PAHs, PBDEs, and organochlorine pesticides such as DDTs, HCHs, and chlordanes were passively sampled using PUF disks in cities of 3 countries (Mexico, Sweden and United Kingdom). Ratios between indoor and outdoor air concentrations (I/O ratios) were estimated because they may be a good tool to indicate whether there are indoor sources (I/O>1) or outdoor sources (I/O<1).

Plants are typically "green" air samplers of natural environmental materials, and hence, they can also be used as passive biosorbent materials to retain semivolatile compounds. A recent study sampled polluted indoor air with burning plastic floor and electronic scrap using SPMDs and fresh unpolluted spruce needles, finding that above mentioned membranes can absorb much more polychlorodibenzodioxins (PCDDs), polychloro-dibenzofurans (PCDFs) and PCBs than spruce needles [28]. In a later study, it was found that spruce needles can complement SPMDs in passive air sampling of PAHs, because the needles tend to uptake more of particle-bound PAHs in comparison to the SPMDs, which mainly sequester PAHs associated with the vapour phase [29]. In the light of using biomaterial as-sorbents of pollutants for environmental biomonitoring, tree bark has been used as a passive sampling medium for atmospheric POPs [30, 31], playing an important role in identifying source/sink regions. This biological sampling medium has unique features, as it simultaneously accumulates both gas-phase and particle-phase POPs from the surrounding air, it can reflect time-integrated overall air pollution levels and conveniently reveal the history of air pollution by the so-called "tree-bark pocket" [32]. Passive adsorption of PAHs by pine bark has studied with the aim of evaluating the air quality of an Italian city [33]. Few investigations have involved ambient air concentrations estimations from their bark contents, generally because there is a lack of mathematical models for bark/air partitioning. Thus, Zhao *et al* presented a mathematical model describing the bark/air partitioning of 50 POPs, including PAHs, organochlorine pesticides, PCBs and brominated flame retardants [32], derived from the comparison of the air measurements using PUF/Tenax/PUF sandwich sorbent tubes with the POP levels in tree bark from 4 different species (pine, poplar, camphor and Chinese fir).

A new promising passive sampler for time-weighted average (TWA) estimations is a needle trap device (NTD), developed and validated by Gong *et al* for BTEX (a mixture of benzene, toluene, ethylbenzene and xylenes) determination [34]. The NTD consists of a needle and a sorbent (0.6 mg Carboxen 1000) positioned at a distance from the tip of the needle, and immobilized by a small quantity of epoxy glue. The sampling process can be described by the Fick´s first law of diffusion, which expresses the relation between TWA concentrations to which the passive sampler is exposed and the mass of analytes adsorbed to the packed sorbent in the sampler. Desorption of target compounds is carried out by thermal desorption, and hence, it becomes a solvent-less procedure. Furthermore, the introduction of the sample using NTD can be automated, facilitating the use of this technology for industrial hygiene applications.

In active sampling, a defined volume of air is pumped through an adsorbent tube packed with one or more adsorbents, where the pollutants are retained at a specific and controlled flow-rate. Active devices such as pumps and flow meters are required to force the flow of the sample through the trap, and to measure the sample volume or the flow rate. When glass or quartz fiber filters are added, particles can also be sampled.

In order to achieve an efficient trapping, it is important to select an appropriate sorbent between the wide variety of materials available in the market. Published in the literature, there are some comprehensive reviews on adsorbent materials suitable for air analysis [7, 35, 36, 18]. The requirements that an adsorbent must fulfil are summarized in table 3. For a better comprehension, a classification of the adsorbents commonly used for air monitoring is shown as well in table 4. When a high adsorption capacity is needed to trap the pollutants, e.g. for sampling very volatile compounds, carbonaceous sorbents, such as activated carbon, the most common sorbent in occupational hygiene determinations for sampling workplace air, carbon molecular sieves, mainly produced by pyrolysis of organic polymers, or graphitized carbon blacks, made out of thermal carbon blacks, are usually the best choice. A recent example is the paper by Vainiotalo *et al*, who sampled 16 VOCs in restaurant air contaminated with environmental tobacco smoke, using sorption tubes filled with Carbopack X, a graphitized carbon black [37]. Problems as irreversible adsorptions, adsorption of water or the formation of artefacts can be encountered derived from their high adsorption capacity and thus, from the high desorption temperature needed when they are used in thermal desorption processes [7]. For example, retention of fluorocarbons, greenhouse gases, is generally high on activated charcoals, but this adsorbent type can cause irreversible retention, possible degradation and is more difficult to use practically due to its heterogeneous composition [38]. However, quantitative recoveries can be also obtained by using strong charcoal-based sorbents like coconut-shell charcoal to sample e.g. compounds as 4-vinyl-1-cyclohexene which is usually found in the air of rubber industry [39]. In the study carried out by Stanetzek *et al*, only activated charcoal was suitable for the efficient sampling of nonpolar highly volatile pollutants, among all the sorbents tested, including Amberlite XAD-4, XAD-8, Tenax GC, Porapak R, silica gel and a modified version of the latter [40].

Table 3. Sorbent requirements for being used in air monitoring (adapted from ref. 36)

Sorbent requirements
Complete adsorption of analytes
Complete desorption of analytes
Fast desorption of analytes
Homogeneous
Chemically inert
No breakthrough
Possibility of long storages
Low background
Trace impurities characterized
No degradation of the sorbent or of the adsorbed analytes
Low affinity to water
Low adsorption capacity for inorganic compounds of air
High mechanical stability
High thermal stability
Re-utilization
Lifespan

Table 4. Sorbents commonly used for air monitoring

SORBENTS BASED ON CARBON	Carbon Molecular Sieves	Carboxens	Carboxen 563 Carboxen 564 Carboxen 569 Carboxen 572 Carboxen 1000 Carboxen 1003 Carboxen 1004 Carboxen 1012 Carboxen 1018
		Carbosieves	Carbosieve SII Carbosieve SIII Carbosieve G
		Carbosphere	
		Anasorb CMS	
		Spherocarb	
	Graphitized Carbon Blacks	Carbotraps	Carbotrap B Carbotrap C Carbotrap F Carbotrap X
		Carbopacks	Carbopack B Carbopack C Carbopack X Carbopack Y
		Carbographs	Carbograph 5
	Activated carbon		
POROUS ORGANIC POLYMERS	Tenax	Tenax TA Tenax GR Tenax GC	
	Chromosorbs	Chromosorb 101-106	
	Porapaks	Porapak P Porapak N Porapak K	
	Haye Sep	Haye Sep P Haye Sep A	
	Amberlite XAD resins	XAD-2, Supelpaks XAD-3 XAD-8	
	PUF		
INORGANIC MATERIALS	Silicagel		
	Zeolithes		
	Alumina		

Trabue *et al* studied the quantification of odorants such as butanoic acid, acetic acid, indoles or phenols in humid environments, and tested different sorbent materials [41]. These compounds have important implications in the area of air emissions from animal agriculture. The carbon molecular sieves performed poorly at high relative humidity due to excessive sorption of water, whereas graphitized carbon sorbent tubes performed best with quantitative recoveries of the tested compounds at all relative humidity. Some tests were also performed sampling different volumes of air, finding Tenax breakthrough of the more volatile odorants, and the tendency to form artefacts with increasing volumes. However, graphitized carbon blacks do not exhibit breakthrough at all the sampling volumes tested. Thus, graphitized carbon blacks were the adsorbents of choice for field samples taken inside swine feeding operations and from a poultry facility. With the purpose of assessing trihalomethanes (THMs) uptake, Caro and Gallego tested Tenax TA, Chromosorb 102 and Carbopack B in humid air samples from an indoor swimming pool and alveolar air samples form swimmers [42]. In terms of adsorption efficiency and breakthrough volume, Chromosorb 102 was the most appropriate sorbent for air sampling. To reduce or avoid the water that is brought into the gas chromatography system via the sorbent materials, a drying tube was placed in front of the sampling tube packed with a hygroscopic salt to reduce water retention at the sampling stage. Na_2SO_4, $CaCl_2$ and K_2CO_3 have been tested for this purpose. Na_2SO_4 has proved to be the most effective, since it was the most resistant to the humid air stream path. In fact, only the 25% of the salt bed has liquated after the passing of 6 L of humid air.

When sampling is carried out in the field, far away from the laboratory where the desorption will be carried out, or e.g. in remote areas (campaigns in ships, polar regions, etc.) where there is not any chromatographic system, it is important to know the storage stability of the target compounds retained on the preconcentration sorbent. Storage stability of organic compounds onto common sorbents will be discussed exhaustively in the performance section of this book. As well, it must be taken into account that, for some applications, especially to evaluate workplace exposures, it is convenient the utilization of light and compact portable devices, to carry out what it is called personal sampling. A recent example is the paper written by van Netten, where a small personal air sampling device is designed to monitor levels of tricresyl phosphate isomers in the cabins of aircrafts [43].

Porous organic polymers comprise a broad range of adsorbent materials differing in the monomer used for polymerization in the production process. One famous porous sorbent is Tenax, which has been extensively used for the air sampling of e.g. VOCs [44-49], musks [50], carbonyls [51], or even less volatile compounds as PCBs [52, 53] or pesticides [54-58]. Its hydrophobic nature is an advantage over some common hydrophilic sorbents as charcoal and silica gel, because air humidity may reduce the sorption efficiency by displacing the organic molecules previously retained [59]. Due to its characteristic high temperature limit (350°C), this sorbent is compatible with solvent and thermal desorption systems, although some authors have noticed changes in Tenax TA from re-used thermal desorption tubes [60]. Tenax can undergo decomposition in highly oxidizing atmospheres generating oxygenated compounds [61, 62]. A few degradation products from Tenax GC, such as benzaldehyde, acetophenone, and higher aldehydes (octanal, decanal) are well known [36, 63]. It has been demonstrated that olefins trapped in polymers as Tenax can react with the ozone causing the formation of interferences. For that reason, several compounds as thiosulfate have been employed to remove ozone before entering into the sorbent trap, and an extensive review has been published on the subject [64]. The use of an ozone scrubber as MnO_2 under certain conditions has also demonstrated to reduce the ozone in the sample stream to near zero and the artefacts observed [65]. Comparing with Tenax GC, Tenax TA has been developed to reduce column bleeding, while Tenax GR, a mixture of Tenax with a 23% of graphitized carbon, provides the possibility of sampling higher volumes of air.

Other common porous polymers are e. g. Amberlite XAD-2 and XAD-4 non-ionic macroreticular resins, used e.g. for collecting the soil fumigant chloropicrin from air [66], or polyurethane foam, widely used for the sampling of pesticides [67-74], PCBs [75], PCDDs/Fs [76], phathalates [77, 78], brominated flame retardants [79-83], organophosphate esters [47], or PFCs [84] in air, but exhibits breakthrough of more volatile compounds [7]. Amberlite XAD-4 and Porapak R were found to be useful for sampling polar and nonpolar low-volatility compounds, whereas Amberlite XAD-8 retained only polar high-melting substances [40]. Regarding inorganic compounds, silica gel was progressively replaced by other sorbents because of its hydrophilic behaviour, and now it is only used for sampling very polar compounds, such as methanol or amines [35, 40]. It is worth a mention the utilization of silica gel coated with derivatization agents to actively collect carbonyls in indoor air, as it is proposed by the US EPA Method TO-11A [85]. As an example, Weng *et al* measured the concentrations of carbonyl compounds in indoor and outdoor air of selected public places, including shopping centres, supermarkets, inter-city bus stations, railway stations, cinemas, and a furniture store, using silica gel coated with 2,4-dinitrophenylhydrazine (DNPH) as sorbent material [86]. The average I/O ratios were greater than 1, which indicates that indoor sources contribute significantly to carbonyls, such as indoor materials and anthropogenic activities. Last year, Kazos *et al* tested Tenax, Flosiril, XAD-2 and silica gel to simultaneously determined chlorothanolil, an organochlorine fungicide classified in the B2 group "probable human carcinogen", and its main metabolite in greenhouse air, finding silica gel the most suitable material [87].

For non volatile and strongly adsorbed compounds, sample recovery is the limiting step. On the opposite, very volatile analytes may pass through the sorbent bed without being trapped and hence, breakthrough volumes must be carefully calculated [7] (see performance section). Multiple-packed sorbent tubes are very useful as they afford the opportunity to collect compounds of a wide volatility range combining different sorbents. The best combination and quantities of each adsorbent that can be used within a preconcentration device shall be previously test in the laboratory or check in the literature. The arrangement of the sorbents must be such that the least volatile compounds are trapped on the weakest sorbent at the front end of the tube, and successively more volatile compounds are trapped by increasingly strong sorbents further down the tube, with the most volatile being trapped at the far end [35]. The sorbent strength is a term used to describe the affinity of sorbents for the target analytes. For example, a stronger sorbent is one which offers greater safe sampling volumes for most/all analytes relative to another, weaker sorbent. Generally speaking, sorbent strength is related to surface area, though there are exceptions to this. As a general rule, sorbents are described as "weak" if their surface area is less than 50 m^2g^{-1} (e.g. Tenax, Carbopack C, etc.) and "strong" if the surface area is around 1000 m^2g^{-1} (includes Spherocarb, Carbosieve S-III, Carboxen 1000, etc.) [88]. A closer characterization of the adsorption strength is given by the specific breakthrough volume (BTV) of model compounds [36]. Typical combinations for trapping VOCs include Tenax with Carboxen [89], Carbopack [90] or Carbotrap [91], Carbopack with Carboxen [92, 93] or Carbosieve [91, 94], and Carbotrap with Carbosieve [95]. XAD-2 resin sandwiched between two polyurethane foam plugs has become one of the most popular traps for less volatile compounds as PAHs [96-100], phthalate esters [96, 101], brominated flame retardants [96, 102] or pesticides [96, 103]. For collecting a wide range of odour nuisance and air-quality VOCs, including alkanes, aromatics, aldehydes, alcohols, chlorides, esters, ketones, terpenes, amides, isocyanates and carbon disulfide in air, another possibility is the combination of Carbotrap, Carbopack X and Carboxen 569 [104]. Supelpak-2, -2B and -2SV are purified Amberlite XAD-2 resins, cleaned to meet US EPA requirements for specific applications as PAH monitoring in indoor air. Together with C_{18}, Supelpak-2 was selected in order to actively collect some pesticides, such as malathion and fenhexamid in air of an experimental greenhouse after their application in a tomato crop [105]. Herrington and Zhang, proposed a method where a dansylhydrazine (DNSH)-coated silica-based-bonded C_{18} sorbent is utilized to actively collect acrolein, a hazardous air pollutant derived from the combustion of wood, fuel, and tobacco, or from the heating of animal fats and vegetable oils [106]. Automated portable samplers combining several sorbents have also been developed. Bechara *et al* designed a new off-line aircraft instrument called AMOVOC for non-methane hydrocarbon measurements [107]. AMOVOC consists of a compact metallic rack, respecting aeronautical standards, divided into two parts: the sampling compartment with multisorbent tubes packed with Carbosieve SIII and Carbopacks B and C, and the automation compartment. This device was constructed to prevent samples from losses and contamination (e.g. contaminants ingress of contaminant by diffusion), which is crucial when considering sampling of trace gases and experimental operations in airport areas. Thus, after sampling, a multiposition valve is automatically switched to the position of "purge tube" to isolate the non-sampled tubes from air.

SPME can be indistinctively used for active sampling using pumps or passive sampling, in which the fiber is retracted a known distance into its needle housing during the sampling period. For example, Wang *et al* estimated the exposure of the strawberries greenhouse workers to pesticides using passive sampling with SPME [108]. It can also be used for a rapid and simple on-site evaluation of pollutants emitted from building materials [109]. SPME is based on partitioning between the polymeric phase of a fiber and the sample matrix. It provides some advantages over traditional extraction methods, integrating both sampling and extraction steps in a single stage, and offering solvent-free operation. In spite of the limited amount of analyte extracted, all is introduced into de gas chromatograph injection port, allowing for good sensitivity, with cost effectiveness and operational simplicity [110]. Furthermore, it is generally independent of moisture, except at very high moistures. SPME quantification is feasible in no-equilibrium conditions, once experimental parameters are held constant, which considerably reduces sampling time [111].

Extraction of analytes by SPME is highly dependent on the nature of the fiber coating. Target analytes are extracted by absorption when a liquid coating as polydimethlsiloxane (PDMS), or polyacrylate (PA), a solid crystalline coating that turns into liquid at desorption temperatures are used. Coatings, including PDMS/divinylbenzene (DVB), Carbowax (CW)/DVB, CW/template resin (TR) or Carboxen (CAR)/PDMS are mixed coatings, in which the primary extracting phase is a porous solid, extracting analytes via adsorption, and trapping the analytes in the active sites of the solid surface of the polymer [112]. By selecting an appropriate fiber, SPME is applicable to a wide range of compounds. Different fiber coatings that can be supplied by the manufacturer are listed in table 5, although new coatings are being continuously proposed. Some examples are ethoxy/PDMS, polyurethane acrylate, 50% phenyl/PDMS [113] or γ-Al$_2$O$_3$-coated fiber, with good selectivity for alkanes and esters [114]. Besides, alternative methods for the preparation of SPME fibers are being suggested, e.g. by treatment with low-temperature nonthermal plasma [115], obtaining a homemade fiber with higher adsorptive capacities than PA fiber for alcohols, or activated carbon fiber for alcohols and BTEX.

Another strategy is to use SPME as a sample pre-concentration device after a sampling step conducted with conventional methods to enhance the sensitivity of the method, collecting an air sample by active sampling using a sorbent [44, 52, 54, 116, 117], by passive sampling [118], or by whole sampling with a canister [119], a bag [120-123], or a glass bulb [124], followed by the extraction of analytes by exposing a SPME fibre. Some examples are the combination of Tedlar bag sampling and SPME for the analysis of trimethylamine, a well-known indicator of spoilage, as it releases rotten seafood odours [123], or for the analysis of volatile sulphur compounds in a poultry factory suffering from sever odour problems [120].

Solid-phase dynamic extraction (SPDE), a novel technique based on the same principles of SPME has been developed by Lipinski [125]. Instead of using a coated fibre, SPDE uses a sorbent coating on the inner wall of a longer stainless steel needle, resulting in a reduced fragility of the needle. As far as we know, Van Durme *et al* have reported the first application of SPDE to the air monitoring, constructing a homemade auto-sampling device to investigate the extraction of toluene from air [126]. They have proposed an accelerated SPDE procedure to avoid dispensing stages during extraction, with a total extraction time of 1.7 min and lower detection limits than those obtained with SPME and conventional headspace sampling with a gas syringe.

Another alternative could be the use of membranes with sorbent interface (MESI), which has been applied to monitor, for instance, plant fragrances emitted into indoor air, by connecting the membrane extraction module made of a piece of hollow silicone fiber to a cryofocusing and thermal desorption sorbent interface [127]. VOCs could also been determined in laboratory air by using a MESI on-line system consisting in a PDMS membrane and two different traps (PDMS and Tenax) [128].

Table 5. Characteristics of commonly used SPME fiber coatings (Supelco, 2009, http://www.sigmaaldrich.com/)

Stationary phase	Film thickness (μm)	pH	Maximum Temperature (°C)	Recommended Operating Temperature (°C)	Application
PDMS	100	2-10	280	200-280	Volatiles
	30	2-11	280	200-280	Non-polar semivolatiles
	7	2-11	340	220-320	Non-polar high molecular weight compounds
PDMS/DVB	65	2-11	270	200-270	Volatiles, amines and nitroaromatic compounds
PA	85	2-11	320	220-310	Polar semivolatiles
CAR/PDMS	75	2-11	320	250-310	Gases and low molecular weight compounds
DVB/CAR/PDMS	50/30	2-11	270	230-270	Flavour compounds, volatiles and semivolatiles or trace compound analysis

Two novel promising passive devices based on low-density polyethylene (LDPE) membrane tubes have also been developed [129, 130]. One of these devices is a spiral-rod sampler consisting of a LDPE membrane acting as a permeation film, and a silicone elastomer as the receiving material; the other is a stir-bar sampler with the same membrane material but a PDMS-coated stir bar acting as the collector phase. The first sampler is cheaper and showed higher sensitivity compared to the second one. Anyway, both samplers have been successfully tested for the long-term air monitoring of SVOCs in a polluted area. Membrane introduction mass spectrometry (MIMS) can be also applied for the on-line detection of SVOCs, including pesticides [131]. The method uses a composite membrane, made by plasma deposition of a thin PDMS layer on a microporous polypropylene support fiber.

Chapter 3

DESORPTION AND SAMPLE TREATMENT

To extract the retained compounds for the analysis, solvent extraction is usually employed, although volatile analytes are normally released by thermal desorption.

Solvent extraction is advantageous for relatively high boiling point or thermolabile analytes; it allows versatility for solvent selection, as well as successive analysis of the same sample. Also, it does not need additional equipment. On the contrary, solvent extraction commonly requires the use of high solvent volumes, which in turn leads to time-consuming sample concentration and purification steps. In addition, many procedures report the contribution of ultrasounds or microwaves to quantitatively extract the trapped compounds. Other techniques, such as supercritical fluid extraction, have been selected to extract the analytes. In this way, volatile organic peroxides, such as tert-butyl perbenzoate or di-tert-butyl peroxide, were extracted by supercritical fluid CO_2 with methanol as modifier, after their collection on Carbotrap and Carbotrap C [132].

Thermal desorption allows the compounds to be quantitatively transferred to the measuring equipment, usually a gas chromatograph, leading to lower detection limits. Since it is solvent-free, it prevents overlapping of the analyte peaks with the solvent or with the peaks of solvent impurities. Nevertheless, it presents some drawbacks to be used in the analysis of emerging pollutants in air. Many of the compounds under this consideration are semivolatile (with relatively high boiling points) and thermally degradable, such as PBDEs. On the other hand, the use of thermally resistant sorbents is a limitation. In fact, only a few sorbents can be used with thermal desorption, e.g. Tenax, graphitized carbons, or Chromosorb. Additionally, this technique only allows performing a single analysis by sample, and a focusing system to avoid the enlargement of the chromatographic band is much recommended. Due to the usual low concentrations of pollutants in indoor air, a preconcentration step prior to their determination is often required in order to reach acceptable detection limits. Besides the routinely applied procedures involving solvent evaporation, some other methods combining extraction techniques have been proposed. Elke *et al* combined a passive air sampling technique using a charcoal pad with two extractive techniques; solvent extraction and SPME, to determine BTEX mixtures in indoor air of buildings, a train or a car [118]. However, the proposed procedure is tedious and time-consuming. Wei *et al* used a combination between microwave-assisted thermal desorption (MAE) and headspace solid-phase microextraction (HS-SPME) to extract PAHs previously collected on a XAD-2 adsorbent [134]. The analytes were desorbed into the extraction solution, evaporated into the headspace by use of microwave irradiation, and absorbed directly on a solid-phase microextraction fiber in the headspace. As it was aforementioned, the combination of solid-phase extraction with Tenax or Florisil and SPME also led to sensitive methodologies for the analysis of many emerging pollutants in indoor air samples [44, 52, 54, 116, 117].

A more detailed description on desorption and sample treatment procedures is given in chapter 6.

Chapter 4

DETERMINATION TECHNIQUES

Determination of contaminants in air is usually performed by gas chromatography using conventional capillary columns (30 m x 0.25 mm I.D., 0.25 µm film thickness) with common stationary phases, such as 5 % phenyl substituted methylpolysiloxane and dimethylpolysiloxane. Mass spectrometry (MS) is the most extended detection technique and it is commonly operated in the electron impact (EI) mode both in the full scan mode and in the selected ion monitoring (SIM) mode [135-137]. Used together with suitable preconcentration techniques, SIM mode allows achieving limits of detection (LODs) in the pg m^{-3} level. Sensitivity and selectivity can be also achieved in the MS-MS mode [72, 138-140]. The negative chemical ionization mode (NCI) [141-143] provides a very high sensitivity and selectivity for brominated compounds mainly when the most abundant fragment, Br$^-$ (m/z= 79/81) is selected.

Many emerging pollutants in air possess one or more halogenated atoms in their structure, which can make electron capture detection (ECD) selective and sensitive enough to analyse them. This is the case of brominated flame retardants [144], and of many pesticides [145-147]. This detector also gives high responses to other electronegative groups as in the case of nitromusks [148].

Other detection techniques, such as nitrogen-phosphorus (NPD) or atomic emission detection (AED) are compatible with several groups of chemicals that can be found in indoor air and thus, particular aspects will be discussed later in this book.

On-line determination techniques do not usually provide very low detection limits, but they can be convenient for atmospheres seriously polluted e.g. emissions from chimneys in industrial applications, where sensitivity is sacrificed to the benefit of rapid analysis. A fully automated instrument combining a continuous wave cavity ring-down spectrometer and dual-trap (using Tenax and Carbosieve SIII) preconcentrator has been recently implemented for monitoring acetylene [149]. This non-methane volatile organic compound is determined because it is emitted into the atmosphere almost exclusively by human activities. The instrument was shown its potential to be deployed in many rural and urban environments, and to be capable of fast analysis with a LOD of 8 pptv without the requirement of calibration.

Although it is not common in air monitoring, in some scarce cases, determinations can also be carried out by spectrophotometric analysis. It is the case of dimethylamine, an odorous and toxic compound because of its role in nitrogen cycling, nutrient transfer and atmospheric acidity. This compound can be determined by sampling with C_{18}-packed solid phase extraction cartridges, derivatization inside the cartridges with a reagent, recovery of the derivatives formed by flushing with a mixture of water and acetonitrile, and final quantification by measuring the absorbance of the collected extracts [150]. This procedure was applied to analyze amines in air samples collected in several fish stands of a city market. Only by visual inspection, it is also possible a semiquantitative estimation by observing the coloured area of the cartridge, and to distinguish between primary and secondary amines. But, as it was obviously expected, the disadvantage of this procedure is its high limit of quantification (LOQ) (2 µg).

It is also worth mentioning the case of carbonyls analysis. The most common method for their determination is to collect them onto DNPH coated solid sorbent followed by solvent extraction and analysis of the derivatives by HPLC/UV, as it is suggested in the US EPA Method TO-11A [151]. Low resolutions between carbonyls with similar properties and the difficulty to identify unknown compounds when an UV detector is used are some of their disadvantages. HPLC/MS-MS can solve some co-elution problems, and greatly improve the LODs, especially for higher molecular weight compounds, but it is an expensive instrument. In comparison with HPLC, gas chromatography improves the resolution and sensitivity, allowing more accurate identification and quantification of carbonyl compounds in complex air samples. The oven temperature programming is generally faster than gradient elution in HPLC, with shorter equilibration times between injections, and produces no disposable waste. Thus, Li *et al* developed a method to determine 20 gaseous carbonyl compounds in the C_1-C_{10} range by GC/MS determination, obtaining LODs ranging from 3.7 to 11.6 ng per sample [51].

Chapter 5

METHOD QUALITY EVALUATION

NIOSH and the Occupational Safety and Health Administration (OSHA) of the United States have published a guideline document to evaluate sampling and analytical methods for airborne contaminants [152]. This document defines some terms related with method performance in air analysis, such as accuracy, storage stability, capacity (breakthrough), sampling rate, recovery, sensitivity, etc. and provides an experimental protocol to define the evaluation criteria to be used for method evaluation. This is of great importance to ensure good traceability of the data obtained and for data comparison at national and international levels. Main parameters related to method quality evaluation are summarized in table 6.

The sampling medium should be studied to evaluate the formation of artefacts, the collection of interferences, or the loss of collection efficiency during storage prior to use. Analysis of blank samples over a period of time can give an indication of whether the sampling medium contributes with artefacts or interferences to the analysis [152]. Blanks of the sampling sorbent or instrumental blanks of the determination system must be periodically run and checked to control interferences and/or contamination. For many compounds, especially for aromatic compounds, sorbent blanks increase with their continued use, but they can be reconditioned back to the original levels or lower. Periodic preconditioning (after every 15-20 uses) for trace-level monitoring is usually sufficient to avoid very high sorbent blank values. Extended sorbent conditioning time invariably leads to lower blanks, but the blank reduction varies from compound to compound [153]. The ubiquity of certain pollutants such as phthalates or synthetic musks, among others, constitutes a severe problem of contamination of the whole analytical system and hence, special attention should be paid to control the blank levels throughout the analysis [154, 155].

Table 6. Main parameters involved in method quality evaluation

Sampling step	
Temperature	Breakthrough (capacity)
Humidity	Flow rate
Ozone variation	Sampling/uptake rate (passive)
Wind speed (passive)	Reverse diffusion (passive)
Rain, snow (passive)	Badge orientation (passive)
Blank of sorbent/sampling step	Storage stability
Collection efficiency	Stability of analytes
Extraction/Desorption efficiency	Degradation of sorbent/analytes
Sample loading	Artefact formation
Determination step	
Instrumental blanks	Instrumental sensitivity
Linearity of the instrumental response	
Complete method	
Recovery	Accuracy
Total blank (whole method)	Repeatability (precision)
Linearity (calibration)	Reproducibility
Evaluation range	Sensitivity (LODs, LOQs)

Breakthrough volume of adsorbents is an important parameter for evaluating the properties of the adsorbent. Target analytes must not exhibit breakthrough in the selected adsorbent to avoid losses during sampling. The term specific breakthrough volume is defined as the volume of gas that causes a compound to migrate through an adsorbent bed of one gram at a specific temperature [156]. It can also be described as the volume sampled (flow rate per time) until mass found on the backup section of the sampler totalled 5% of the mass found on the front section of the sampler [152]. Therefore, it is considered that no losses occur when the breakthrough volume is less than a 5% [35].

Breakthrough volume can be determined by indirect and direct methods. In the indirect method the sorbent tube is a chromatographic column and the breakthrough volume is calculated from the measurement of chromatographic retention volumes for the sorbent at different temperatures. The direct methods include the passage of a constant concentration of the analyte through the sorbent trap connected to an appropriate detector, the field sampling with dual traps in series and loading traps by liquid or vapour deposition, followed by purging with a volume of gas and determining the amount of sample remaining [157]. The best method is that which is close to the real sampling procedure.

It is known that there are several factors influencing the breakthrough volumes of compounds adsorbed on sample tubes. Factors include concentration of analytes, temperature of adsorption, humidity, and the amount of sorbent; another factor is the presence of other compounds that may be alike or differ greatly in their properties (polarity, volatility, etc.) [158]. Different authors have measured breakthrough volumes at different flow rates, finding no significant differences, and proving that the breakthrough volume is almost independent of flow rate [157, 159, 160]. Those parameters which can influence the breakthrough volume are summarized in table 7. These many factors affecting breakthrough volumes cannot be simulated entirely in the laboratory. Thus, it is recommended to reduce tabulated or measured data by at least one-third to ensure a quantitative sampling [36].

Table 7. Parameters affecting the breakthrough volume

Parameter	Explanation
Analyte concentration (sample loading)	BTVs can decrease if the capacity of the adsorbent is exhausted
Temperature of adsorption	Adsorption is an exothermic process
Humidity	Affect if the adsorbent retains noticeable amounts of water
Amount of sorbent	Higher amount of sorbent leads to an increase of the BTVs
Sampling flow	It must allow a sufficient time for the interactions between analyte and adsorbent surface. BTVs appear to be independent of the flow-rate
Presence of other compounds	Can decrease the BTVs (displacement of target analytes by molecules with a higher affinity by the sorbent)

As it was commented in the sampling section of this book, it is important to know the storage stability of the target analytes in the sorbent or in the medium used for their collection. If analytes are decomposed into the device used for their collection during the sampling step or if some losses occur while they are retained in the adsorbent before being analyzed, the recovery of the compounds will decrease, leading to non-quantitative recoveries. Obviously, this is particularly important in the case of samples that must be taken far away from the laboratory where the analysis will be carried out. Sample stability is defined as the ability to retrieve the analyte from the sampler after storage for a period of time under a defined set of environmental conditions [152]. In addition, it must be taken into consideration that compounds strongly retained may not be further extracted quantitatively. Thus, this concept is deeply related to the recovery of the analyte from the medium, which is a quality parameter that must be determined. Many studies have shown that storage on adsorbents, particularly for very volatile compounds, can be improved by keeping the enriched devices at sub-ambient temperatures. However, it must also be kept in mind that uptake of ambient contaminants can be greater at lower temperatures [161].

The storage stability of organic compounds on several sorbents has been documented. Storage stability of various ketones on different activated carbons and the effect of adsorbed water vapour under different storage conditions, including temperature, have been evaluated [162]. The effect of water must be taken into account mostly in the sampling of VOCs. Their retention on adsorbents changes and the safe sampling volume is reduced when water exists. In addition, difficulties would also be caused in the chromatographic system, even irreversible damage to the capillary column. Single-walled carbon nanotubes, a novel adsorbent for collection VOCs avoid the reduction of recovery due to the presence of water, simplifying sampling devices and experimental procedures [163]. This is possible because of the hydrophobic and nonpolar suface of this new sorbent, containing an analog-graphite hexatomic ring structure. It is convenient to notice that water vapour is always present in the air and its adsorption on the surface of sorbents based on carbon is a serious drawback mainly for polar compounds. It was found that the performance of synthetic carbons was better than for coconut charcoals because the former contain less inorganic impurities and have a lower capacity for water adsorption. The water adsorption and the ash content of the carbons can be a measure of the reactive sites that may chemisorb ketones or catalyze their decomposition, suggesting in this way that both parameters can act as an indicator of their storage stability on activated carbons.

Regarding storage temperatures, better recoveries may be obtained when samples are stored at 4°C. Storage stability of VOCs on graphitized carbon blacks (Carbograph 2 and 5) was studied, showing no losses after one week, but obtaining severe losses for polar compounds such as alcohols and ketones after a period of one month of storage [161]. The stability of Chromosorb 106, Carboxen 569, Tenax GR and Carbosieve S-III charged with chlorinated and non-chlorinated hydrocarbons has been studied over a period of months and years, showing especially Tenax good results [158]. The stability of 52 VOCs, including alkanes, alkenes, aromatics and terpenes was studied too on Carbosieve S-III and two Carbotraps, obtaining recovery values close to 100% for all compounds after 1-week sample storage, and thus, proving perfect analyte stability on the multilayer adsorbent bed [16].

Estimated recovery is another quality parameter that should be calculated. It is defined as the ability to recover and determine an analyte placed in or on a sampler. It is estimated by dividing the amount recovered from a sampler by the amount of analyte fortified in or onto the sampler. The recovery of the analyte from the sampler should be ≥ 75%. If recovery varies with analyte loading, results should be graphed as recovery vs. loading, so that appropriate correction can be made to sample results [152].

Recovery problems can indicate a strong retention of the analytes by the sorbent, losses and/or degradation of the analytes during sampling, extraction or determination stages. Certain compounds can not be sufficiently collected because they decompose by daylight during air sampling even in an indoor environment, obtaining very low recoveries and collection efficiencies. Ozone or nitrogen oxides present in the air can react with the alkene groups of molecules, causing their losses during sampling [164]. This can occur for example for some pesticides like allethrin, tetramethrin, phenothrin or cyphenothrin, which are photodegraded relatively quickly due to their photolability [137, 165-167]. Organoiodine fungicides, such as 3-iodo-2-propynyl-N-butylcarbamate (IPBC), 1-bromo-3-ethoxycarbonyloxy-1,2-diiodo-1-propene (BECDIP) or 1-(4-chlorophenyl)-3-iodopropargylformal (CPIP) are also relatively susceptible to photoirradiation and decomposed easily through photochemical reactions [137]. Different ozone scrubbers such as sodium thiosulfate were investigated to remove oxidizing compounds before entering in the sorbent cartridge. Filters or sorbents used for sampling must be directly impregnated with solutions of these compounds to overcome these problems [164, 168, 169]. Nevertheless, excessive recovery rates of even up to 150% can sometimes be obtained. This situation can be found when a high-level blank is obtained, or when interferences co-elute at the same retention time. Very high recoveries can be calculated as well by the matrix enhancement effect, more precisely described as matrix-induced chromatographic response enhancement. It is explained by the blocking of active sites in the injection port due to sample matrix, resulting in reduced analyte loss during injection and higher response values of the measured components. Elflein *et al* have overcome this problem using matrix-matched calibration when sampling household insecticides with a glass fibre filter and two PUF plugs [170].

Repeatability and reproducibility should be also estimated. IUPAC gold book defines repeatability as the closeness of agreement between independent results obtained with the same method on identical test material, under the same conditions (operator, apparatus, laboratory and after short intervals of time), and reproducibility as the closeness of agreement between independent results obtained with the same method on identical test material but under different conditions (operators, apparatus, laboratories and/or after different intervals of time) [170]. Both are usually reported as a relative standard deviation or as a variation coefficient.

For air analysis, calibration is generally performed by using gas conditioned in pressure cylinders at various concentrations covering the range of interest [171, 172]. Gas standards can be used for preparation of calibration curves; certified gases for some analytes are available from suppliers with NBS (National Bureau of Standards) traceability documentation. For calibration, gaseous standard mixtures were normally prepared by syringe injection of standard solutions of the target compounds into flasks or canisters previously evacuated with a pump. The syringe injection method is particularly recommended when calibration for numerous compounds in a mixture is required. Dynamic generation of standard gas mixtures can also be utilized carrying the components contained in diffusion tubes in a small stream of nitrogen to a gas dilution section. Thus, target compounds permeate through a convenient membrane into a complementary gas flow.

When adsorbents are used for sampling, external calibration is feasible. Therefore, a viable solution is direct spiking of standard mixtures of the target compounds in an appropriate solvent on the sorbent. This calibration technique was used, for example to determine PCBs [52, 54], pesticides [53, 173] or synthetic musks [50] in air. Then, the spike is left to homogenize with the adsorbent, in suitable conditions depending on the tested compound, e.g. at low temperatures for several hours. When volatiles are sampling, an air sample polluted with the compounds of interest can be simulated. Thus, a drop of a solution of volatile analytes in a solvent can be carefully placed in a V-shaped glass tube that it is connected to the sorbent cartridge. After that, a selected volume of air can be pumped throughout the system until the air is enriched in the analytes before reaching the sorbent [116, 117, 44, 147]. A calibration can be carried out by placing small volumes of liquid standard solutions of known concentration in the glass tube. A schematic diagram of the sampling device used for the calibration of chlorobenzenes following this procedure is shown in figure 1.

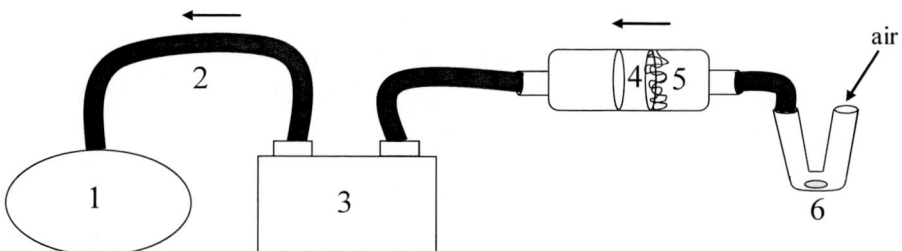

1: vacuum pump, 2: PTFE connectors, 3: flow meter, 4: Tenax TA, 5: glass wool, 6: V-shaped glass tube containing the analytes.

Figure 1. Schematic diagram of the air-sampling device used for calibration in the analysis of chlorobenzenes in indoor air (from ref. 44).

Calibration problems can be encountered e.g. when actively sampling SVOCs using quartz fiber filter disk combined with an Empore disk [137]. Calibration curves for flame retardants as tris(2-butoxyethyl)phosphate or tris(2-ethylhexyl)phosphate, insecticides like cyfluthrin, deltamethrin, imiprothrin, prallethrin, fenitrothion, furamethrin or pyridaphenthion, and fungicides like BECDIP, CPIP, IPBC, 2-(thiocyanomethylthio)benzothiazole (TCMTB), cyproconazol or thiuram were found to be quadratic functions. The detected quantities of these compounds tended to decrease at low concentrations. On the other hand, the curves for the insecticide hydramethylnon or the fungicides dichlofluanid or difolatan were sigmoid. Linearity problems for some of these calibration curves have been attributed to several factors like photochemical decomposition, adsorption in the injection port of the gas chromatograph or in the chromatographic column, the need of derivatization, or inexactnesses when GC/MS is utilized [174].

The development of passive sampling techniques will be accelerated if a good calibration method is achieved. The most widely used method is the calibration based on equilibrium extraction. It only reflects the analyte concentrations at the time that the samplers are retrieved, and it requires known distribution coefficients between the extraction phase and the sample matrix. Recent proposed calibration techniques include the in-fibre standardization and standard-free kinetic calibration, both pre-equilibrium approaches which require known distribution coefficients. For passive samplers with long equilibrium times, the calibration can be performed in the linear range, and the sampling rates should be previously determined in the laboratory or predicted with empirical equations [17].

Calibration is the main obstacle for applying SPME to indoor air monitoring. Nevertheless, great efforts have been made to overcome this problem [175-181]. For example, volatiles in an air stream were quantified by Bartlet and Zilkowski by SPME in a dynamic mode and under non-equilibrium conditions, slowly evaporating the analytes from a rubber septum [178, 179]. Mangani and Cenciarini exposed a Carbograph coated fibre to a flask, where permeation tubes containing the target analytes were placed while a stream of nitrogen flowed through the system [182]. Especially for pesticides that have very low vapour pressures, generating gas-phase calibration samples of defined concentration is a difficult task to solve. For assessing the linearity of SPME determination of various pesticides, Ferrari *et al* developed an assembly for the headspace extraction of analytes from vials containing spiked solutions of pesticide mixtures in water [183]. The flask contains a glass-coated stirring bar and it is immersed in a thermostatic bath. Air saturated by pesticide vapours is dynamically pumped through a glass flask where a SPME fibre is inserted by means of a rubber septum. Therefore, calibration is performed from a vapour-saturated air sample.

Sensitivity of the analytical method should also be estimated by calculating the LODs and the LOQs. LOD is the lowest mass that can be detected. It is defined as the mass of analyte which gives a mean signal $3\sigma_b$ above the mean blank signal, where σ_b is the standard deviation of the blank signal. LOQ is the lowest mass that can be quantified; this is that can be reported with acceptable precision. It is the largest of the mass corresponding to the mean blank signal + 10 σ_b, or the mass above which recovery is ≥75% [152]. LOD and LOQ can be determined as the concentration when signal to noise ratio (S/N) is 3 or 10, respectively. Recovery, repeatability and sensitivity of recent developed methods for indoor air analysis of selected contaminants have recently been reviewed [184, 185].

Chapter 6

SELECTED APPLICATIONS

6.1. FLAME RETARDANTS

Brominated Compounds

The widespread use of plastics and other synthetic materials in electrical appliances, construction materials and textiles has increased the flammability of these products and led to the extensive use of flame retardants to improve their flame resistance and to meet the fire safety standards. Measurements conducted in workplaces as offices, internet cafes, computer rooms and computers or electronic shops indicated significantly higher levels of brominated flame retardants compared to furniture stores, homes and outdoor air [186].

In the last years brominated compounds used as flame retardants have attracted great attention as emergent contaminants [187]. Most brominated flame retardants have been found to bioaccumulate in biota and humans [188]. Their widespread production and use, together with the increasing contamination of the environment, wildlife and people, highlights the importance of identifying emerging issues associated with the use of brominated flame retardants, especially in indoor environments. The most used brominated flame retardants are polybrominated diphenyl ethers (PBDEs), hexabromocyclododecane (HBCD), tetrabromobisphenol-A (TBBPA) and polybrominated biphenyls (PBBs). Other less known compounds like bis(2,4,6-tribromophenoxy)ethane (BTBPE) and decabromodiphenyl ethane (DeBDethane) have an increasing interest due to their emerging use as substitutes of octaBDE and decaBDE commercial mixtures, respectively [81].

Polybrominated diphenyl ethers constitute an important class of brominated flame retardants commonly added to a variety of consumer products. Their production began in the 1960s and they were incorporated in building materials, electronic equipment, lighting, electric wiring, textiles, furniture, floor coverings, industrial paints, and in many other common products. Due to their persistent and bioaccumulative nature, penta- and octabrominated commercial mixtures have been banned within the European Union and their use in North America has recently begun to be phased out [189]. However, vast reservoirs of PBDEs remain in existing consumer products, potentially contributing to environmental and human burdens of PBDEs for decades [190]. PBDEs are incorporated into materials as additives and thus they may be released into air through volatilization during the product lifetime and, as a consequence, levels are expected to be elevated in indoor air.

Intake through food consumption is an undoubtedly important source of human exposure to PBDEs. However, the potential for exposure to PBDEs in the indoor environments is also real as inhalation and inadvertent ingestion of contaminated dust have been reported to be the largest contributors of PBDEs exposure of toddlers through to adults [191]. In addition, and because of higher concentrations, indoor air and dust likely represents a significant source to outdoor air [192].

The analysis of some brominated flame retardants, such as TBBPA, HBCDs and the higher brominated PBDEs, is a relatively new challenge for most analytical laboratories. Special emphasis must be given to the need of an adequate Quality Assurance /Quality Control (QA/QC) protocol, which is necessary for the reliable analysis of these environmental contaminants at trace levels [193]. Literature on the analysis of brominated flame retardants in different matrices, paying special attention to new analytical developments and quality assurance requirements, has recently been reviewed by Covaci *et al* [194].

Polybrominated diphenyl ethers can be expected in any laboratory environment equipped with computers and other electronic devices. Significant concentrations of BDE47 and BDE99 have been identified in laboratory air by Thomsem *et al* [195]. Thus, in order to avoid a high content of brominated flame retardants in the procedural blanks, it is important that all materials involved in the sample preparation are properly cleaned, and that direct exposure of the sample to the laboratory air is minimized. A proper glassware cleaning implies a thermal treatment at 450 °C and solvent rinsing before being used. Polyurethane foam sorbents are usually precleaned by Soxhlet extraction with different solvents prior to sampling step. Moreover, the use of plastics should be reduced as possible in the determination of brominated flame retardants, since they can contain a wide range of these compounds. For the same reason, unnecessary electric appliances and upholstered furniture should be avoided as well as unpackaging of goods in the laboratory where extraction and clean-up take place.

Of special interest and concern is BDE209, the primary component in the decaBDE commercial mixture—actually the most important PBDE mix in production. This compound, as well as other highly brominated congeners, is photosensitive, so direct exposure to UV light should be avoided. Thus, incoming sunlight into the laboratory, as well as possible UV light from fluorescent tubes should be blocked by means of UV filters. Herrmann *et al* [196] reported up to 70% decomposition of BDE209 when stored for 24 h under light conditions. Wrapping glassware with aluminium foil during sample treatment, and using amber glassware are simpler preventive measures to minimize UV-degradation of the analytes. Additional recommendations regarding this issue can be found in de Boer and Wells [197].

Sampling and Analysis

Sampling of brominated flame retardants in indoor air and particulate matter usually implies an active procedure (see Table 8). In general, sample volumes ranging from a few hundreds of litres to less than 30 m^3 are enough to reach indoor LODs in the low ng m^{-3} level for most compounds. However, lower limits have been reported for sample volumes between 100 and 385 m^3 [80, 84, 201]. Polyurethane foam is the most used sorbent for sampling brominated flame retardants in indoor air [80, 84,141-143, 199, 201, 202], although XAD-2 resin has also been employed [81, 82, 200]. Rudel *et al* [96] used this resin sandwiched between two polyurethane foam plugs for sampling volumes from 10 to 14 m^3 at flow rates between 8-9 L min^{-1}. Other active systems for indoor sampling are based on the use of solid-phase extraction (SPE) disks or cartridges of materials such as styrene-divinylbenzene [32].

Table 8. Analytical procedures for the determination of flame retardants in indoor air

Analytes	Sampling sorbent	Desorption and sample treatment	Determination	Recovery (%)	LOD	Year	Ref.
Tri-DecaBDE, BTBPE, DeBDethane	Cellulose pad and XAD-2 (25 m^3, flow 50 L min^{-1})	Soxhlet with toluene previous addition of ^{13}C-labeled surrogates. Clean-up on treated silica (KOH+H$_2$SO$_4$), elution with hexane, clean-up on a GPC-system, elution with hexane/dichloromethane (1:1), concentration to a small volume and addition of tetradecane and ^{13}C-labeled IS	GC/NCI-MS (SIM)	12-97	2.30-173 pg m^{-3}	2002	81
Tri-DecaBDE	PUF (0.4-0.7 m^3 min^{-1}, indoor: 175-385 m^3)	Soxhlet with acetone/hexane (1:1), previous addition of ^{13}C-labeled surrogates. Addition of activated copper, concentration to a small volume and clean-up on an acid/basic multilayer silica column, concentration to 0.2 mL and addition of ^{13}C-labeled IS	GC/NCI-MS (SIM)	74-87	0.28-28.6 pg m^{-3}	2002	198
Tetra-PentaBDEs	XAD-2 and PUF (10-14 m^3, 8-9 L min^{-1})	Soxhlet with 150 mL hexane/diethyl eter (6%) containing a deuterated surrogates (p-terphenyl-d$_{14}$). Concentration to 2 mL and addition of deuterated IS (PAHs)	GC/EI-MS (SIM)	40-220	NR	2003	96
Di-DecaBDE, 2,4,6-TBPh, PBPh, HBB, HBCD	Empore C18 (14.4 m^3, flow 10 L min^{-1})	Ultrasound extraction with 10 mL acetone. Concentration of 5 mL of extract to 0.5 mL and addition of deuterated IS	GC/AED	81-91	0.47-9.9 ng m^{-3}	2003	199

Analytes	Sampling sorbent	Desorption and sample treatment	Determination	Recovery (%)	LOD	Year	Ref.
Tri-HeptaBDEs, TBBPA, 2,4,6-TBPh	1) Passive adsorption on glass funnel surface. 2) Isolute ENV+ (0.18 m^3, flow 4 L min^{-1})	Elution with 6 mL dichloromethane/methanol (7:3). Concentration to 30 µL, derivatization with 50µL diazomethane and addition of IS (TBB)	GC/NCI-MS	NR	NR	2004	200
Tetra-DecaBDE, BTBPE, BB-209, TBBPA	1) PUF (1.5 m^3, flow 3 L min^{-1}) 2) Cellulose pad and PUF (3.6 m^3, flow 9 L min^{-1})	Ultrasound extraction with 5 mL dichloromethane, addition of surrogates (BDE-128, TrBCBPA), solvent exchange to hexane, concentration to 0.1 mL and liquid extraction with 2 mL methanolic KOH ($\geq$ 50 %) twice: 1) Neutral fraction (hexane): clean-up on a silica/ H_2SO_4 (2:1) column and elution with 8 mL hexane 2) Aqueous fraction: acidification with HCl and liquid extraction with hexane/MTBE (1:1). Separation of the organic phase, concentration to 1 mL and derivatization of phenolic compounds with 0.2 mL diazomethane. Clean-up on a silica/H_2SO_4 (2:1) column and elution with 8 mL dichloromethane	GC/NCI-MS (SIM)	~97 (23-60 for BTBPE, TBBPA)	LOQ: 3-100 pg m^{-3}	2004	203
Tri-HeptaBDEs	PUF disk (42 days, uptake rate 2.5 m^3 day^{-1}: 105 m^3)	Soxhlet with dichloromethane/hexane (1:1), previous addition of surrogates (BDE-35, BDE-181). Concentration and solvent exchange to hexane, clean-up on silica/alumina (2:1), elution with 100 mL hexane/dichloromethane (1:1), concentration and solvent exchange to dodecane and addition of IS (Mirex)	GC/NCI-MS (SIM)	80-90	0.2-0.5 pg m^{-3}	2004	84

Table 8. (Continued).

Analytes	Sampling sorbent	Desorption and sample treatment	Determination	Recovery (%)	LOD	Year	Ref.
Tri-PentaBDEs	PUF disk (50 days, uptake rate 1.12-1.95 m^3 day^{-1}: 56-98 m^3), previous addition of surrogates (PCB-19, PCB-147)	Soxhlet with 200 mL hexane, previous addition of ^{13}C-labeled surrogates. Concentration to 2 mL, treatment with 2 mL $H_2SO_{4(c)}$, liquid extraction with dimethylsulfoxide, clean-up on florisil, elution with 20 mL hexane. Concentration and solvent exchange to 20 µL nonane and addition of IS (PCB-29, PCB-129)	GC/EI-MS (SIM)	42-80	NR	2004	143
TBBPA	PUF (3 m^3, 3 L min^{-1})	Ultrasound extraction with 5 mL acetonitrile, previous addition of ^{13}C-labeled surrogates. Concentration to 0.5 mL, filtration through a syringe filter, elution with 5 mL methanol, concentration to 0.1 mL and addition of 0.075 mL water	LC/ESI-MS (SIM)	75-107	NR	2004	210
Tri-DecaBDE, BTBPE, DeBDethane	Cellulose pad and XAD-2 (1 m^3, flow 2 L min^{-1})	Soxhlet with 250 mL toluene, previous addition of ^{13}C-labeled surrogates. Concentration to 0.5 mL, clean-up on treated silica (KOH+H_2SO_4) and GPC, solvent exchange to nonane and concentration to 40 µL	GC/NCI-MS (SIM)	NR	0.01-1.3 ng m^{-3}	2006	204

Analytes	Sampling sorbent	Desorption and sample treatment	Determination	Recovery (%)	LOD	Year	Ref.
Tri-HeptaBDEs	Passive using organic films from window surfaces (exposition < 4 months) collected using kimwipes wetted with isopropyl alcohol	Soxhlet with toluene/acetone (4:1), previous addition of ^{13}C-labeled surrogates. Acid-base washing (H$_2$SO$_4$ and KOH), concentration to dryness, reconstitution in dichloromethane/hexane (1:1). Clean-up on a multilayer acidic/basic silica column, elution with dichloromethane/hexane (1:1), and solvent exchange to hexane. Clean-up on a copper column, elution with hexane, clean-up on alumina, elution with dichloromethane/hexane (1:1), concentration to dryness and reconstitution in toluene	GC/EI-MS (SIM)	35-119	NR	2006	205
Tetra-HeptaBDEs	PUF (100-200 m^3, flow 0.4 m^3 min^{-1})	Soxhlet with petroleum eter/acetone (1:1). Concentration to 1 mL, solvent exchange to isooctane, addition of ^{13}C-labeled surrogates, clean-up on a multilayer (basic, neutral, acidic, neutral) silica column and elution with 60 mL dichloromethane/hexane (1:1). Clean-up on alumina, elution with 60 mL dichloromethane/hexane (1:1), concentration to < 10 mL and addition of ^{13}C-labeled IS	GC/EI-MS	> 98	0.3-20 pg m^{-3}	2006	202

Table 8. (Continued).

Analytes	Sampling sorbent	Desorption and sample treatment	Determination	Recovery (%)	LOD	Year	Ref.
Tetra-HexaBDEs	PUF (300 m^3 flow 0.6-0.8 m^3 min^{-1})	Soxhlet with dichloromethane/hexane (1:1), previous addition of ^{13}C-labeled surrogates. Treatment with $H_2SO_{4(c)}$, clean-up on acidic silica, elution with hexane, clean-up on florisil, concentration and solvent exchange to nonane	GC/EI-MS (SIM)	54-104	1 pg m^{-3}	2007	207
Tri-HexaBDEs	PUF disk (21 days, uptake rate 2.5 m^3 day^{-1}; 50 m^3) previous addition of surrogates (BDE-3, d6-γ-HCH, PCB-107, PCB-198)	Soxhlet with petroleum eter, previous addition of surrogates (BDE-2, BDE-35). Concentration to 0.5 mL, solvent exchange to isooctane and addition of IS (Mirex)	GC/NCI-MS (SIM)	110-116	1.2-18 pg m^{-3}	2007	79
Tri-HexaPBDEs	PUF disk (28 days, uptake rate 1.1-1.9 m^3 day^{-1}: 31-53 m^3), previous addition of surrogates (PCB19, PCB147)	Soxhlet with hexane, previous addition of ^{13}C-labeled surrogates. Concentration to 2 mL, treatment with 2 mL $H_2SO_{4(c)}$, liquid extraction with dimethylsulfoxide, clean-up on florisil, elution with 20 mL hexane. Concentration and solvent exchange to 20 µL nonane and addition of IS (PCB-29, PCB-129)	GC/EI-MS (SIM)	45-67	0.1 pg m^{-3}	2007	82
Tri-DecaBDE	PUF (9 m^3, 2 L min^{-1})	PSE extraction with petroleum eter, previous addition of ^{13}C-labeled surrogates. Concentration to 0.2 mL and filtration through glass wool	GC/NCI-MS (SIM)	NR	NR	2007	22

Analytes	Sampling sorbent	Desorption and sample treatment	Determination	Recovery (%)	LOD	Year	Ref.
HCDBCO	PUF disk (21 days, uptake rate 2.5 m^3 day^{-1}: 52.5 m^3)	Soxhlet with petroleum eter. Concentration to 0.5 mL and solvent exchange to isooctane	GC/NCI-MS (SIM)	NR	1.3 pg m^{-3}	2007	144
Tri-DecaBDE	XAD-2 (6-26 m^3, flow 13-18 L min^{-1})	Soxhlet with dichloromethane. Concentration to a small volume, clean-up on silica, elution with 50 mL dichloromethane, concentration to 0.3-0.5 mL, solvent exchange to isooctane and addition of IS	GC/NCI-MS (SIM)	64-90	NR	2008	80
Mono-DecaBDE	PUF (2 m^3, flow 3 L min^{-1})	Ultrasound extraction with 5 mL dichloromethane, previous addition of ^{13}C-labeled surrogates. Concentration to 1 mL, solvent exchange to hexane, concentration to 1 mL, clean-up on a Isolute NH_2 cartridge	GC/NCI-MS (SIM)	NR	NR	2008	23
TiBP, TBP, TCEP, TCPP, TPhP, TBEP, TEHP	PUF (2.1 m^3, flow 3.0 L min^{-1})	Ultrasound extraction with 5 mL dichloromethane, previous addition of surrogate (TPP). Filtration through glass wool, concentration to a small volume and addition of IS (ABP)	GC/NPD	>95	0.1 ng m^{-3}	2000	215
TPhP, IPPDPP, PPDPP, TBPDPP, TBP, TCEP, TCPP, TBEP	PUF (1.5 m^3, 3 L min^{-1} and (3.6 m^3, 9 L min^{-1})	Ultrasound extraction with 5 mL dichloromethane previous addition of surrogate (MDPP). Concentration to 0.1 mL	GC/NPD	> 95	NR	2001	141
TCEP, TCPP	PUF (1 m^3, 5 L min^{-1})	Soxhlet with hexane/acetone (4:1). Concentration to small volume	GC/EI-MS (SIM)	NR	1 ng m^{-3} (LOQ)	2001	135

Table 8. (Continued).

Analytes	Sampling sorbent	Desorption and sample treatment	Determination	Recovery (%)	LOD	Year	Ref.
TBP, TCEP, TPhP, TBEP, TEHP, TCrP, TCPP, TDCPP	PUF (1.4-3.4 m³, flow 4 L min⁻¹)	Ultrasound extraction with 37 mL dichlorometane, previous addition of surrogate (TPP). Solvent exchange to hexane, concentration to 0.1 mL and addition of IS (Phenanthrene-d₁₀)	GC/EI-MS	62-100	0.073-0.41 ng m⁻³	2004	212
TEP, TPP, TiPP, TiBP, TCEP, TCPP	Non-equilibrium SPME (100 μm PDMS, 60 min) with controlled linear airflow (7cm s⁻¹)	Thermal desorption (2 min, 250 °C)	GC/NPD	NR	~2 ng m⁻³	2004	218
TEP, TPP, TiBP, TBP, TCEP, TCPP	Equilibrium SPME (7 μm PDMS 12 h or 100 μm PDMS 24 h) with controlled linear airflow (10 cm s⁻¹)	Thermal desorption (2 min, 250 °C)	GC/NPD	NR	7 μm PDMS 0.1 ng m⁻³ 100 μm PDMS 0.01 ng m⁻³	2004	219
TMP, TEP, TPP, TiPP, TiBP, TBP, TCEP, TCPP, TPhP, TTP	Cellulose filter (1.4 m³, flow 3 L min⁻¹) previous addition of surrogate (MDPhP)	Ultrasound extraction with dichloromethane. Concentration to a small volume	GC/PCI-MS-MS	NR	0.1-1.4 ng m⁻³	2004	140
TCrP, TEP, TPhP, TPP, TBEP, TCEP, TDCPP, TEHP	Empore C18 (7.2 m³, flow 5 L min⁻¹)	Ultrasound extraction with 8 mL acetone, and shaking. Centrifugation, decantation of 5 mL supernatant, addition of IS (fluoranthene-d₁₀) and concentration to 0.3 mL	GC/EI MS (SIM)	94-112	0.1-0.6 ng m⁻³	2004	137

Analytes	Sampling sorbent	Desorption and sample treatment	Determination	Recovery (%)	LOD	Year	Ref.
TMP, TPP, TBP, TCPP, TCEP, TDCPP,TPhP, TBEP, TEHP, DOPP, TEEdP, CLP1	Isolute NH_2 (1.0-2.7 m^3, flow 2.5 L min^{-1})	Elution with 10 mL dichloromethane, previous addition of surrogate (TPeP). Concentration to dryness, dissolution in dichloroethane and concentration to 0.1 mL	GC/NPD	82-110 (34-58 TEEdP, TMP, TPhP)	0.1-3.9 ng m^{-3}	2005	216
TEP, TPP, TiBP, TCEP, TCPP	Non-equilibrium SPME (100 μm PDMS, 40-90 min) or equilibrium SPME (30 μm PDMS, >18 h) with controlled linear airflow (10-35 cm s^{-1}: flow 1.1-3.8 L min^{-1})	Thermal desorption (2 min, 250 °C)	GC/NPD	NR	NR	2005	220
TEP, TiPP, TPP, TBP, TCEP, TCPP, TDCPP, TBEP, TPhP, DPEHP, TEHP, TTP	Isolute NH_2 (1.5 m^3, flow 2.5-3.3 L min^{-1})	Elution with 5 mL methyl-tert-butyl eter, previous addition of SS (THP). Addition of IS (TPeP)	GC/NPD	~ 100 %	0.1-0.3 ng m^{-3}	2005	217
TMP, TEP, TPP, TBP, TCPP, TCEP, TEHP, TBEP, TDCPP, TPhP, TCrP	Empore C18 (14.4 m^3, flow 10 L min^{-1})	Ultrasound extraction with 10 mL acetone. Concentration of 5 mL of extract to 0.5 mL and addition of IS (tris(1H,1H,5H-octafluoropentyl) phosphate)	GC/FPD	90-100	0.24-3.5 ng m^{-3}	2007	207

Passive air samplers based on polyurethane foam disks are being increasingly employed for sampling of brominated compounds in indoor air [22, 23, 203-205]. They are particularly attractive because of their facility to obtain time-integrated samples in indoor locations, where active samplers would not be practical over long time periods. In passive sampling, conversion of contaminant masses per sample into concentrations in air requires knowledge of the air uptake rate of the disk samplers and their deployment time. Wilford *et al* [203] estimated an average uptake rate of 2.5 m^3 per day for tri- to hexaBDES. Sampling time usually ranges between 20 and 50 days, which approximately yields to air volumes from 50 to 100 m^3.

Another approach is the use of organic films from window surfaces as time-integrated passive samplers for PBDEs [206]. These organic films are formed by condensation of gas phase species and organic aerosols as well as by deposition of particulate-associated compounds. With knowledge of the uptake rate and film-air partition coefficient (K_{FA}), it is possible to estimate gas-phase air concentrations assuming that compounds in film and the gas-phase in air are at equilibrium.

Brominated flame retardants are commonly extracted from sorbents by Soxhlet extraction which, despite its drawbacks, is still widely used due to its general robustness and high extraction efficiency. In this way, recoveries higher than 98 % for tetra- to heptaBDEs after Soxhlet extraction with dichloromethane and petroleum ether/acetone (1:1) have been reported [84]. Ultrasound-assisted extraction can be advantageous for the extraction of PBDEs and other brominated compounds [141, 143, 199, 202, 207] since this technique allows shorten extraction times and uses smaller solvent volumes. Very recently, a pressurized solvent extraction (PSE)-based procedure was applied by Allen *et al* [79] for the analysis of tri- to decaBDE in residential indoor air. Glass fibre filters and polyurethane foam plugs were extracted separately with dichloromethane and petroleum ether respectively. Extractions were completed in 5 min and, although higher costs were initially involved compared to Soxhlet extraction, the reduced extraction time and lower solvent consumption decreased the long-term cost and made the PSE more environmentally friendly. After extraction, a variety of clean-up procedures on silica gel, alumina, florisil or combinations of these sorbents are commonly used to improve the sensitivity for further analysis [22, 201, 204]. A complete procedure is described by Karlsson *et al* [81], who pre-cleaned Soxhlet extracts on a KOH/H$_2$SO$_4$-treated silica column followed by a clean-up on a gel permeation chromatography (GPC) system before analysis with GC/MS. Recoveries, evaluated by addition of ^{13}C-labeled surrogate standards, were in the range from 12 to 97% for tri- to decaBDE, and LODs lower than 0.2 ng m^{-3}.

Separation of brominated flame retardants is generally performed by means of GC/MS. Nevertheless, thermal degradation during their chromatographic separation has been reported for highly substituted PBDEs, mainly BDE-209. Degradation of these compounds leads to low repeatability in their analysis [198] and thus, special attention must be paid to ensure a proper GC analysis. Residence time in the column has been shown to be critical. If the residence time is too long, thermal degradation of highly substituted congeners, especially BDE-209, is substantial. Shorter standard columns were initially used for the analysis of decaBDE and so, the analysis of PBDEs, both the low and the high brominated congeners, required the use of two columns of different length. Although this approach is still in use in many laboratories, the development of narrow bore columns has allowed a proper determination of all congeners using only one column [208]. Narrow bore columns, with maximum length 8–10 m, small internal diameter (0.10 mm), and a thin film coating (0.10 µm), achieve comparable resolution in shorter analysis times [208, 209]. A comprehensive study on the influence of main GC parameters on the determination of decaBDE has been performed by Bjorklund *et al* [210]. According to these authors, the on-column injector is the most suitable injector for clean samples analysis, whereas programmable temperature vaporizing (PTV) injector provides a good compromise between robustness and yields for more complex samples.

A further optimization of GC analysis of the highly substituted PBDEs has recently been described by Regueiro *et al* [144]. This study has been focused not only on decaBDE separation but also on the octa- and nona-brominated ethers, obtaining satisfactory results in terms of yield, accuracy and precision using a narrow bore column and a split/splitless injector operated at 320 ºC.

Mass spectrometry is the most widely used detection system for analyzing brominated flame retardants in indoor air samples, specially operating in the negative chemical ionization mode (NCI) [23, 80-82, 141-143, 195, 199 ,200, 203, 205]. This technique provides a very high sensitivity and selectivity for brominated compounds, especially with selected ion monitoring of the most abundant fragment, Br$^-$ (m/z= 79/81). However, there may be problems with identification and co-elution of other brominated compounds, and it is not possible the use of ^{13}C-labelled compounds as internal surrogate standards (SSs) [202]. Using GC/NCI-MS, Gevao *et al* [205] determined tri- to heptaBDEs in indoor air reaching LODs from 0.2 to 0.5 pg m^{-3}. Mass spectrometry in the EI (SIM) mode has also been employed for quantification of this kind of compounds in indoor air [22, 84, 96, 201, 204, 206], reporting LODs in the range 0.3-20 pg m^{-3} for the analysis of tetra- to hexaBDEs [84]. The presence of Br atoms in the molecules of the compounds allows the use of gas chromatography with an atomic emission detector (AED). In this way, a wavelength of 827 nm was selected for Br detection and LODs in the low ng m^{-3} were obtained for most compounds in indoor air. [207].

Analysis of TBBPA and 2,4,6-tribromophenol —used as BFR and also the major breakdown product of TBBPA— by GC requires a previous derivatization step. In this way, acetylation was carried out with diazomethane [141, 195]. The use of LC/MS in the determination of TBBPA is an alternative that provides several different detection modes and eliminates the need of derivatization. For the determination of TBBPA in air, Tollback *et al* [202] developed a LC/MS method using electrospray ionization (ESI) in the negative ionization mode with SIM. This kind of ionization was compared to atmospheric pressure ionization (APCI), achieving LODs between 30-fold and 40-fold lower.

Concentration in Indoor Air

Several studies have reported concentration levels of brominated flame retardants in air from electronics recycling facilities [96, 141]. Sjodin *et al* [141] investigated the presence of several of these compounds in an electronics recycling plant and other indoor work environments in Sweden. The highest concentrations of all the identified brominated flame retardants were found in the recycling facility. For the rest of sampling sites, the corresponding concentrations in air were, in general, several orders of magnitude lower. Most abundant compounds in the recycling plant were BDE183, BDE209, BTBPE, and TBBPA with mean values in the range 19-36 ng m^{-3}. On the other hand, BDE-47 was the most abundant PBDE congener in a computer teaching hall and a circuit board assembly plant with a mean concentration of 0.76 and 0.35 ng m^{-3}, respectively.

Indoor air concentrations of tri- to hexaBDEs have been measured in homes in Canada, detecting up to 1600 pg m^{-3} [203]. These values were higher than those reported in indoor air in Kuwait, with an average concentration in homes of 15 pg m^{-3} [205]. Shoeib *et al* [84] determined concentrations in homes ranging between 76 pg m^{-3} and 2088 pg m^{-3} for tri- to heptaBDEs, whereas those reported by Chen *et al* [80] were in the range 0.3-1710 pg m^{-3}. In Figure 2, the levels found in several homes and one office in different countries are depicted. Harrad *et al* [201] reported levels of tetra- to hexaBDEs in outdoor and indoor air from different microenvironments including offices and homes. Concentrations of the tetra- and pentaBDEs in indoor air were always higher than those detected in outdoor air. Values for all studied compounds ranged from <1 pg m^{-3} to 1330 pg m^{-3}. Indoor air concentrations of brominated flame retardants were generally higher in offices than in homes [201, 203, 204, 207]. A correlation between the concentration of several PBDEs and the number of electrical appliances and polyurethane foam-containing chairs in sampled rooms was observed [201]. Several studies have also conducted the analysis of laboratory air [84, 195], showing the presence of PBDEs in the pg m^{-3} level. All these studies point out the ubiquity of these types of contaminants.

Recently, high levels of PBDEs have been measured in car interiors [186], estimating a daily intake inhalation of 0.5-2909 pg day^{-1}, with a contribution of 29% to the overall daily exposure, comparable with residential exposure. This study demonstrates that car interiors can be an important route of human exposure to PBDEs via inhalation.

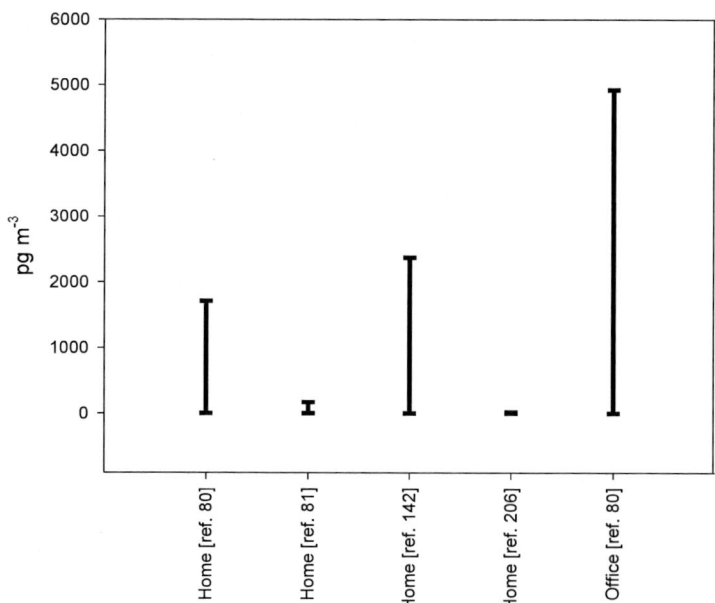

Figure 2. Concentration levels of several brominated flame retardants in indoor environments.

Organophosphate Esters

Organophosphate esters (OPs) are manufactured on a large scale to be used as flame retarding agents and/or plasticizers in a variety of products such as electronic equipment, lubricants, plastics, glues, varnishes and furnishing fabrics. Several studies demonstrated the potential of these materials to emit phosphate flame retardants as well as their degradation products [207, 211]. OPs may diffuse out at rates depending on their vapour pressures and the ambient temperature, and are thus emitted to the surrounding air [212]. Consequently, there are abundant sources of OPs in both public and domestic buildings, including diverse building materials and consumer products. Several toxicological effects of organophosphate triesters have been reported, although very little is known about their health impact on humans. However, some reviews indicate that a number of these compounds, for instance tri-n-butyl phosphate (TBP), tris(2-chloroethyl) phosphate (TCEP) and tris(2-chloropropyl) phosphate (TCPP), may negatively affect human health [213, 214]. As well as for brominated flame retardants, indoor environment represents the main source of human exposure to these pollutants through inhalation of air and inadvertent ingestion of dust. The most volatile OPs are found in the gas phase, whereas the OPs with higher molecular mass are mainly associated to the suspended particulate matter and dust [141, 212].

Sampling and Analysis

Organophosphate flame retardants have been mainly collected from indoor air and particulate matter by active sampling (see Table 4). Sample volumes between 1 and 14 m^3 are usually employed at flow rates ranging from 1 to 10 L min^{-1}. Most of sampling devices consist on a glass or quartz fibre filter for collecting the particulate matter, and one or several polyurethane foam plugs for the gas phase [141, 211, 215]. Saito *et al* [207] and Yoshida *et al* [137] described active sampling methods for organophosphate compounds in air using a quartz fibre filter disk followed by a C_{18} SPE disk. The main advantage of this disk-type configuration is the lower restriction of the flow rate. The use of aminopropyl silica SPE cartridges has also been proposed as a simple alternative for collecting both the gas phase and the particulate matter [216, 217].

Air sampling using SPME has mostly been applied to more volatile compounds than organophosphate flame retardants. As it is known, semi-volatile compounds diffuse more slowly than VOCs, and thus require longer sampling periods to reach their air/fibre partition equilibrium. However, a dynamic air sampling method based on SPME was developed by Isetun et al [218-220], in which a controlled linear air flow is generated over the fibre in order to increase agitation and thus minimize the static layer surrounding the fibre. As a result, an increase in the extraction rate is produced and consequently the equilibration time is shortened. Extracted compounds are almost entirely from the gaseous phase, so no information about contribution of airborne particulate matter is obtained. Organophosphates are present primarily in the particle-associated phase rather than in the gaseous phase. Carlsson et al [215] observed that OP esters were mainly recovered from the filter while the part passing into the polyurethane foam plugs was less than 1%.

Ultrasound-assisted extraction is the most widely used technique for recovering OP compounds from filters and sorbents [221]. Sjodin et al [141] carried out the extraction with 5 mL dichloromethane for 20 min in an ultrasonic bath (power 50 W, frequency 48 KHz). The extraction procedure was repeated once using fresh solvent and recoveries higher than 95 % were obtained after concentration to 0.1 mL. Soxhlet extraction has also been applied by Ingerowski et al [135] with n-hexane/acetone (4:1) for 8 h. In the case of sampling with SPE cartridges [216, 217], extraction can be performed by elution or fractionation with a suitable solvent. Staaf et al [217] extracted organophosphate triesters from aminopropyl silica cartridges by using 5 mL methyl tert-buthyl ether (MTBE) reaching quantitative recoveries.

The use of very selective and sensitive detectors such as nitrogen phosphorus detector (NPD), allows a simple extract preparation, which usually consists on a filtration step followed by concentration to a small extract volume prior to the analysis by GC [211, 215, 216]. In spite of it, LODs in the level of low ng m^{-3} are achieved for most of reported methods.

Organophosphate flame retardants in indoor air have been mainly determined by gas chromatography. In most of cases, NPD is the selected technique for their quantification due to its high selectivity and sensitivity for this kind of compounds. Carlsson et al [211, 215] achieved LODs lower than 0.1 ng m^{-3} with no further extract preparation than filtration through glass wool followed by volume concentration. However, NPD does not offer the possibility for positive identification, so mass spectrometry is sometimes required for confirmation [141, 211, 215, 216].

Mass spectrometry in the EI mode with SIM has also been employed [135, 137, 212] for quantification. Hartmann et al [212] determined OP flame retardants and plasticizers in indoor air obtaining LODs from 0.073 to 0.41 ng m^{-3}. Positive chemical ionization (PCI)-MS-MS in the selected-reaction monitoring (SRM) mode has been applied by Bjorklund *et al* [140] in indoor air samples. A comparative study was performed between EI-MS and PCI-MS-MS under identical sampling and extraction conditions. LODs utilizing GC/PCI-MS-MS were found to be in the range 0.1-1.4 ng m^{-3}, which is about 50-fold lower than those obtained with GC/EI-SIM.

Recently, Saito *et al* [207] have used a flame photometric detector (FPD) for determination of organophosphate flame retardants indoors. This detector presents some of the advantages of the NPD such as high selectivity for phosphorous compounds. LODs between 0.24 and 3.5 ng m^{-3} were achieved with this detection technique.

The selection of suitable surrogate and internal standards (ISs) is conditioned by the extensive use of NPD and the impossibility of using isotopically labelled compounds with this detector. Several compounds such as tripropyl phosphate (TPP) and tripentyl phosphate (TPeP) are among the most frequently used ISs.

Concentration in Indoor Air

Organophosphate flame retardants have been found in the indoor air of a number of homes [140, 207, 216, 219] with concentrations ranging from less than 1 ng m^{-3} up to several μg m^{-3} (see Figure 3).

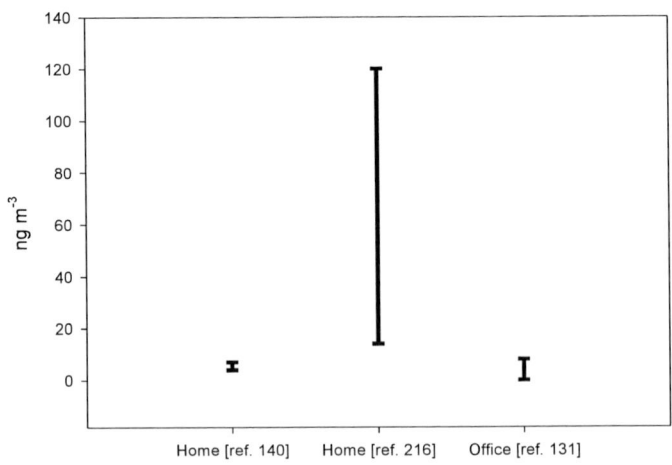

Figure 3. Concentration levels of tributyl phosphate in indoor environments.

In addition to homes, the presence of these compounds in other indoor environments such as offices, public buildings and domestic establishments has been reported [216]. The chlorinated OPs TCEP and TCPP were the most abundant, and were present in all the sampled environments at concentrations up to 730 ng m^{-3}, and 570 ng m^{-3}, respectively. OPs were also found in schools and an office building [215], in which TCEP was in the range 11-250 ng m^{-3}, whereas TBP was present at concentrations from 17 ng m^{-3} to 35 ng m^{-3}; concentrations of triphenyl phosphate (TPhP) were below 0.7 ng m^{-3}, which may be attributed to a lesser migration rate because of its lower volatility. Hartman *et al* [212] determined OPs in several workplaces, e.g. public buildings and cars at concentrations up to 56 ng m^{-3} (for TCEP) and 29 ng m^{-3} (for TBP); TPhP levels were generally lower than 1 ng m^{-3}.

6.2. PLASTICIZERS

Phthalate Esters

Phthalate esters are extensively used as softeners in the production of polymeric materials such as polyvynylchloride (PVC). Since phthalate esters are not chemically bound to the polymer, they can be easily released into the environment. PVC and other polymers are widely produced for building materials and thus, the surrounding environment can be polluted by phthalates. Due to their high volume production and their widespread use, phthalates, as well as some other chemicals present in the domestic environment, are potentially important indoor contaminants. In addition, people working in industrial plants producing plasticizers or living near such plants may be exposed, via indoor air inhalation, to levels of these pollutants that could constitute a significant contribution to the total daily intake [174].

Due to their ubiquity, phthalates can be found everywhere, including common laboratory equipment and reagents. In consequence, the main problem in phthalate analysis is external contamination coming from the sampling and sample preparation procedure and even the chromatographic analysis. This problem has been extensively studied by Frankhauser and Grob [222]. The analysis of blanks is of great importance, as are all the precautions in the treatment of the material and reagents used in any step of the analytical process. To minimize contamination [223, 224], the use of plastic materials should be avoided, the sample preparation procedure should be as simple as possible with minimal extraction steps, and minimal glassware used.

Glassware should be properly cleaned by solvent rinsing and thermal treatment at 400°C. Prior to use, it should be rinsed with blank tested organic solvent (cyclohexane or isooctane) to deactivate its surface. Organic solvents and laboratory grade water usually contain traces of phthalates, even the ones commonly available for trace analysis, and these must be checked to establish background levels. Also, reagents need to be checked. Additional contamination of material, water solvents, and reagents can occur due to the laboratory air. The material should be stored in a closed container or wrapped in aluminium foil to avoid adsorption of phthalates from the air. As previously commented, phthalates can be present in the chromatographic system. The most important contamination source is located in the inlet and gas supply system, inlet septa, liners and o-rings. Since the caps for autosampler vials also contain phthalates, as general precaution, only one injection should be made from each vial [225].

Sampling and Analysis

In Table 9, details on the analysis of phthalate esters in indoor air samples are illustrated. The analysis of atmospheric levels of phthalates would however require high sample volumes of up to 1000 m^3 [228]. But for the analysis of indoor environments, air volumes of 1 to 10 m^3 are usually sampled to achieve low detection limits at the low ng m^{-3} level [99, 174]. Procedures involving only 15 L of indoor air samples have also been reported [229]. The sorbent material employed to retain phthalate esters can be polyurethane foam, Tenax GR, PDMS on Chromosorb, C_{18}, charcoal, or combinations of various sorbents like polyurethane foam and XAD resin. Sorbents are usually pre-extracted using different solvents or solvent mixtures to prevent contamination [99].

Desorption of phthalate esters from sorbent cartridges can be performed by extraction with organic solvents or by thermal desorption (TD). In the first case, extracts are usually concentrated to achieve sufficient overall method sensitivity or for solvent exchanging for further analysis. Before concentration, the addition of anhydrous sodium sulphate avoids the presence of residual water traces in the organic extracts. Either a gentle stream of nitrogen or Kuderna–Danish (K-D) can be used for the concentration of the extracts [230]. Cleanup procedures including the use of fuming concentrated sulphuric acid or silica gel columns [228, 230, 231] have been reported.

Table 9. Analytical procedures for the analysis of plasticizers in indoor air

Analytes	Sampling sorbent	Desorption and sample treatment	Determination	Recovery (%)	LOD	Year	Ref
DEP, DBP, BBP, DEHP	Charcoal granules (1L min^{-1}, 3 days, 4.3 m^3)	Ultrasound with 1 mL toluene and centrifugation	GC/EI-MS (SIM), GC/FPD	97.5-115	25.6-118.6 ng m^{-3}	2001	174
DEP, BBP, DBP, DEHP, DHP, DAMP, DPP, DCHP, DIBP	XAD-2 and PUF (3.8 L min^{-1}, 0.29-5.9 m^3)	Soxhlet with 200 mL of 6% diethyl ether/hexane, addition of sodium sulphate and concentration to 1 mL 10% diethyl ether/hexane. Silylation.	GC/EI-MS (SIM)	95-129 (DEP)	0.0045-1.64 µg per extract (BBP, present in the blanks)	2001	99
DEP, DBP, BBP, DEHP, DCHP, DPP, DIBP	XAD-2 and PUF plugs (8-9 L min^{-1}, 10-14 m^3)	Soxhlet with 150 mL of 6% diethyl ether/hexane, addition of sodium sulphate and concentration to 2 mL 10% diethyl ether/hexane	GC/EI-MS (SIM)	40-220	2-75 ng m^{-3}	2003	96
DMP, DEP, DPP, DBP, DIBP, DCHP, BBP, DEHP, DOP	PUF (5 L min^{-1}, 2 m^3)	PSE with hexane/diethyl ether (95:5), concentration.	GC/EI-MS (SIM)	91-100	Determination limits: 10 ng m^{-3}	2004	77
DBP, DEHP	SepPak PS (2 L min^{-1}, 20–24 h).	5 mL acetone, and concentration to 5 mL.	GC/EI-MS (SIM)	100-102	100 ng m^{-3} (2.88 m^3 air sample)	2004	154
DEP, DBP	Tenax GR (200 ml min^{-1}, 0.1 m^3).	TD (290°C, 10 min) to a cold trap (-30°C) followed by TD (325°C, 15 min)	GC/EI-MS	94-96	5 ng m^{-3}	2005	226
DBP, DEHP	5% PDMS on Chromosorb, (500 mL min^{-1}, 15–30 min, 15 L).	TD (300 °C, 10 min)	GC/MS	NR	1 ng m^{-3}	2006	229

Table 9. (Continued).

Analytes	Sampling sorbent	Desorption and sample treatment	Determination	Recovery (%)	LOD	Year	Ref
Phthalates	PUF, 24 h.	NR	NR	NR	NR	2007	78
DBTC, TBTC, DPTC, TPTC	Activated carbon-fibre filter (7.2 m^3, 5 L min^{-1})	Ultrasound with 10 mL HCl/methanol and benzene. Centrifugation, washing with 15 mL NaCl (10%), drying over Na$_2$SO$_4$ and concentration to 1 mL. Derivatization with propylmagnesium. Addition of 10 mL H$_2$SO$_4$ (0.5 M), addition of 10 mL methanol, liquid extraction with 2.5 mL hexane (x2), and concentration to 0.5 mL	GC/FPD	95-99	0.2-0.4 ng m^{-3}	1993	227

In a leading study on the presence of phthalate esters in the Swedish atmosphere, phthalates retained into polyurethane filters connected in series were extracted with acetone-hexane in an ultrasonic bath [231]. Otake *et al* [174] extracted the phthalates from charcoal tubes by sonication with 1 ml of toluene, obtaining good extraction efficiency (98 to 115%). These authors proved that sonication times longer than 10 min did not improve the results. The use of PSE was described by Fromme *et al* [77] as part of a general procedure to determine phthalate esters and musk compounds in indoor air. In this way, compounds are sampled in polyurethane foam cartridges (2 m^3 air collected), desorbed using 5% dichloromethane in hexane and the concentrated extracts analyzed by GC/MS, achieving determination limits of 10 ng m^{-3}. The use of ultrasounds and pressurized solvents are very advantageous from a sustainable chemistry point of view, but Soxhlet apparatus is still preferred by many laboratories. An extraction procedure of phthalates from a quartz filter and PUF-XAD sampling cartridges using 200 mL of 6% ether in hexane for 16 h was described, achieving LODs of 2-75 ng m^{-3} [96].

As it was mentioned in the section 3, thermal desorption presents some advantages over the solvent-based extraction methods, which are generally derived of the absence of solvent manipulation. In addition, all the retained compounds are thermally desorbed into the GC and hence, high sensitivity can be achieved. However, the high temperatures needed for quantitative desorption of the less volatile phthalates from typical sorbents, such as Tenax or carbon materials, limits the application of this desorption technique. An alternative to overcome this problem could be the substitution of these sorbents by silicones, and thermal desorption-GC/MS [224]. An estimation of the LODs was achieved sampling 15 L air ranged between 1 and 10 ng m^{-3}.

Phthalate diesters are sufficiently volatile and thermally stable to be analyzed by gas chromatography [232]. Although several detector types have been applied to phthalate GC analysis in environmental samples, most of the recently proposed methods involve the use of MSD working in the EI mode [225]. Most phthalates fragment with characteristic ions, such as m/z 149. This is the case of diethyl phthalate (DEP), dibutyl phthalate (DBP), butylbenzyl phthalate (BBP), bis-(2-ethylhexyl) phthalate (DEHP), and diisobutyl phthalate DIBP). Dimethyl phthalate (DMP) fragment with m/z 163, and diisononyl and diisodecyl phthalates (DINP, DIDP) with m/z 307. These fragmentation patterns allow a very sensitive and selective detection, particularly when operating in the SIM mode [77, 99, 96, 154, 174]. Separation columns are usually 25 to 30 m x 0.25 to 0.32 mm I.D. coated with phenyl methylpolysiloxane or dimethylpolysiloxane stationary phases, which allow programming separations in a wide range of temperatures (typically, from about 50 to 300°C, at 10 °C min^{-1}) with low bleeding. As commented above, the ubiquity of phthalate esters constitutes a very real problem through the analysis process, requiring a careful check for blank concentrations for which values of >100 ng m^{-3} have been reported [154, 174].

Concentration in Indoor Air

Phthalate indoor concentrations highly depend on the building materials and the type of furniture at each sampling emplacement. Hence, a broad range of values have been reported for the analyzed compounds (see Figure 4).

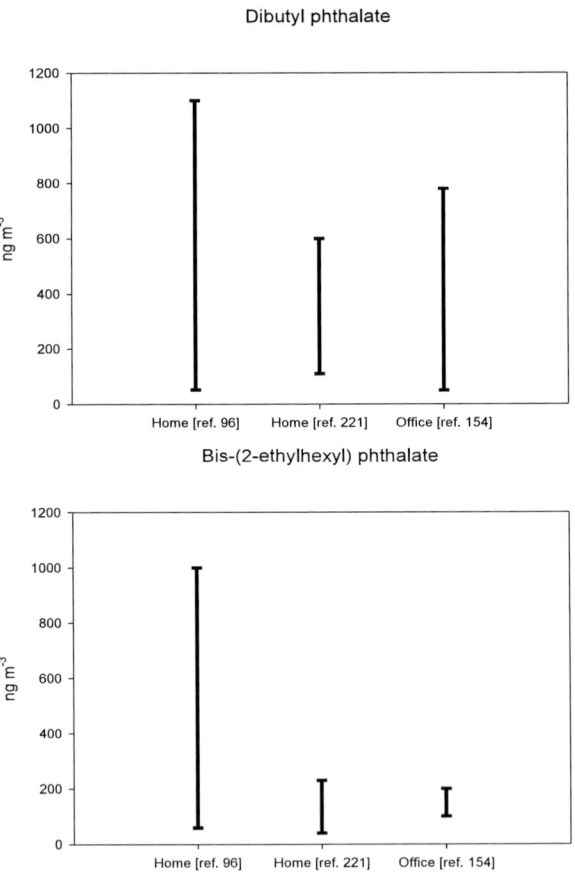

Figure 4. Concentration levels of dibutyl phthalate and bis-(2-ethylhexyl) phthalate in indoor environments.

Results on the monitoring of phthalate esters in 125 homes in California (USA), showed a clear predominance of DBP and DEP in indoor air, with mean values of 410 and 350 ng m^{-3}, respectively [233]. In this study, DEHP (110 ng m^{-3}) and BBP (35 ng m^{-3}) were also found. Higher concentrations of total phthalates (>1000 ng m^{-3}) have been quantified in apartments and homes [77, 154, 174], which demonstrates that DBP predominates in the gas phase of domestic indoor environments. Fromme *et al* [77] extended the study of the indoor occurrence of phthalates and musk compounds to kindergartens, finding mainly DMP and DBP at similar mean concentrations (1100-1200 ng m^{-3}). DBP and DEHP have also been quantified in office rooms, finding concentrations in the broad range found in homes [154]. The indoor exposure to EDCs was studied by Rudel *et al* [96], reporting that phthalates were the most abundant of the 89 organic chemicals considered in the 120 homes surveyed; total concentrations of DEP, DBP, DEHP, and BBP ranged from <90 to 7000 ng m^{-3}, indicating that the sources of these chemicals must be located indoors and highlighting the importance of indoor environments in the total exposure to chemicals.

Organotin Compounds

Organotin compounds are widely employed as stabilizers of polyvinyl chloride (PVC) polymers and as industrial catalysts for polyurethane and silicone elastomers. Hence, they are present in water pipes, food packing materials, polyurethane foams and many other consumer products [227]. The prominent toxicological feature of the organotins is their immunotoxicity, an effect produced by di- and trialkyltins as well as triphenyltins. Furthermore, the importance of organotins as environmental endocrine disrupters and their potential to adversely affect human health, has prompted the European Commission to identify tributyl tin (TBT) as a priority hazardous substance [234].

Organotin compounds have been collected from indoor air by active sampling through quartz filters and an activated carbon-fibre filter [227]. A flow rate of 5 L min^{-1} was employed for 24 h periods, which yields to air volumes of approximately 7 m^3. Extraction was performed by sonication twice with 10 mL 1M HCl in methanol for 10 min, and then twice with 2.5 mL benzene for 10 min. After derivatization with propyl magnesium and several clean-up steps, organotin compounds were analyzed by gas chromatography with FPD. Recoveries higher than 95 %, and LODs in the range 0.2-0.4 ng m^{-3}

were obtained. Measured concentrations of triphenyltin chloride (TPTC) ranged between 0.4 ng m^{-3} and 0.6 ng m^{-3}.

6.3. SYNTHETIC MUSK FRAGRANCES

Synthetic musk fragrances are added in large amounts to toiletries, cosmetics, household products, and a wide variety of other consumer products. Synthetic fragrances such as air fresheners are used in products to scent the environment. There are environmental concerns as synthetic musks contribute to both air and water pollution [235]. They have been measured in workplaces and other crowded indoor environments, although there is an important lack of information about their concentration levels in domestic indoor air. Owing to their chemical structures, synthetic musks can roughly be classified in two main categories: nitromusks and polycyclic musks. Among them, the polycyclic musks Galaxolide (1,3,4,6,7,8-hexahydro-4,6,6,7,8,8- hexamethyl-cyclopenta-(g) 2- benzopyrane, HHCB) and Tonalide (7-acetyl-1,1,3,4,4,6-hexamethyl-tetraline, AHTN) are used in the highest quantities, being the latter included in the US EPA high production volume chemical list [236]. In 1997, the nitromusks musk xylene (1-tert-butyl-3, 5-dimethyl-2, 4, 6-trinitrobenzene, MX) and musk ketone (4-tert-butyl-3,5-dinitro-2,6-dimethylacetophenone, MK) were added to the list of chemicals for priority action of the EU, and in 1998 MX was added to the corresponding list of the Oslo and Paris Commission [235].

In spite of their vast use and exposure, there is limited information available related to their health effects. In addition, physical-chemical properties of synthetic fragrances have more in common with hydrophobic and semivolatile organic pollutants that are known to biomagnify through the food chain [237]. Fragrances can impact indoor air quality and there is suggestive evidence that may play an important role in respiratory diseases and long-term impact [238]. Hence, synthetic musks present enough properties which make them worth considering as a group of indoor air pollutants.

Sampling and Analysis

In the few studies reporting the analysis of musk compounds in indoor air and particulate matter, synthetic musks have been collected by active sampling of 2 to 100 m^3 at, in general, reduced flow rates [77, 136, 155] (see Table 10).

The most typical sorbent for sampling this kind of compounds in the gas phase is polyurethane foam, whereas a glass fibre filter is usually used to collect the airborne particulate matter [136, 155, 239].

Desorption of musk compounds from sorbents has been carried out by Soxhlet using different solvent mixtures such as n-hexane/diethyl ether (9:1) [136, 155] or dichloromethane [239]. PSE has also been used for extraction of musk fragrances from polyurethane foam sorbents with satisfactory yields (recovery = 91-100 %) [77]. Very recently, Regueiro et al [50] applied for the first time the SPME as an alternative to solvent extraction in the analysis of synthetic musks in air including polycyclic and nitro musks. In a firts step, musk compounds are adsorbed onto an amount of only 25 mg Tenax placed in a glass SPE device. After addition of 100 µL acetone to the sorbent to favour desorption, analytes are transferred to a DVB/CAR/PDMS fiber in the headspace (HS) mode. By GC/MS analysis in the full scan mode, detection limits in the pg m^{-3} level were achieved for an air sample volume of 5 m^3. This is a very clean procedure which does not require any other sample treatment. However, the use of certain sorbents in combination with high solvent volumes makes necessary a clean-up procedure performed on silica gel [136, 155] or with a combination of silica gel and alumina [239]. Elution volumes higher than 50 mL solvent are usual thus requiring further concentration of the extracts.

Determination of synthetic musk fragrances is usually performed by gas chromatography using conventional capillary columns (30 m x 0.25 mm I.D., 0.25 µm film thickness), with common stationary phases, including 5 % phenyl substituted methylpolysiloxane and dimethylpolysiloxane. Mass spectrometry is the most extended detection technique for musk compounds and it is commonly operated in the EI mode with SIM [77, 136, 155], which leads to LODs in the pg m^{-3} level. However, nitromusk compounds have also been analyzed in the NCI mode [136, 155] achieving LODs between 100-fold and 60-fold lower with regards to EI mode.

Table 10. Analytical procedures for the determination of synthetic musks in indoor air

Analytes	Sampling	Desorption and sample treatment	Determination	Recovery (%)	LOD	Year	Ref.
HHCB, AHTN, ATII, ADBI, AHMI, DPMI, MX, MK	PUF (2 m^3, flow 5 L min^{-1})	PSE with hexane/diethyl eter (19:1), previous addition of deuterated surrogates. Concentration to a small volume	GC/EI-MS (SIM)	91-100	10 ng m^{-3}	2004	77
HHCB, AHTN, ATII, MX, MK	PUF (36-108 m^3, 25-38 L min^{-1})	Soxhlet with 300 mL hexane/diethyl eter (9:1), previous addition of deuterated surrogates. Concentration to 0.5 mL, clean-up on silica, elution with 50 mL of hexane/ethyl acetate (9:1), concentration to 0.2 mL, and addition of IS (TCN)	GC/EI-MS (SIM), GC/NCI-MS (SIM) for nitromusks	69-126	EI Polycyclic (5-45 pg m^{-3}) NCI: Nitromusk (4-12 pg m^{-3})	2004	136
HHCB, AHTN, ATII, ADBI, AHMI,DPMI	PUF (72 m^3, 0.3-0.4 m^3 min^{-1})	Soxhlet with dichloromethane. Concentration, clean-up on silica-alumina (2:1), elution with dichloromethane, solvent exchange to hexane, concentration to 0.2-0.5 mL, and addition of IS (HMB)	GC/EI-MS	57-107	60-120 pg m^{-3}	2007	239
HHCB, AHTN, ATII, ADBI, AHMI, DPMI, MX, MK, MM	Tenax (1-10 m^3, 100 L min^{-1})	Addition of 100 μL acetone followed by HS-SPME (DVB/CAR/PDMS fiber, (30 min, 100°C)	GC/EI-MS (ITD)	85-103	29-380 pg m^{-3}	2009	50

Deuterated musk xylene and AHTN standards are commercially available for use as surrogate and ISs. Nevertheless, deuterated AHTN has been reported to undergo partial deuterium to hydrogen exchange during analysis, which may result in an inaccurate surrogate recovery [240]. A variety of other surrogate and ISs have also been used in different environmental matrices such as deuterated PAHs, pentachloronitrobenzene, hexamethylbenzene and various labelled and unlabelled PCBs. LODs, repeatabilities and recoveries reported in the analysis of musk compounds in air are summarized in Table 10.

As in the case of phthalates, special care must be taken to reduce the risk of contamination through the analysis due to the extensive presence of musk fragrances in household products, soaps, perfumes and all kind of cosmetics [155].

Concentration in Indoor Air

Kallenborn *et al* [155] reported atmospheric concentrations of nitromusks and polycyclic musks in Norwegian air samples, not only in urban areas but also in remote areas. In one indoor laboratory air sample analyzed during the same sampling campaign, concentrations up to 2.5 ng m^{-3} of HHCB were measured, finding i.e. 10-fold higher than detected in outdoor air, which raises the suspicion that air as a transport and transfer medium for synthetic musks is still underestimated.

Musk compounds were further studied in several indoor workplace environments [136]. Highest values were found in a hairdresser facility with 44 ng m^{-3} HHCB, although a coffee bar contained also high synthetic musk burden with 35 ng m^{-3} of HHCB and 12 ng m^{-3} of AHTN, respectively. The presence of synthetic musk fragrances was also evaluated in indoor air samples from kindergartens in Berlin [77]. HHCB gave the highest levels ranging from 15 to 299 ng m^{-3}, whereas AHTN and Phantolide (6-acetyl-1,1,2,3,3,5-hexamethyl-indane, AHMI) where found at average concentrations of 47 and 22 ng m^{-3}, respectively. Polycyclic musk fragrances were measured in a typical cosmetic plant and surroundings, finding concentrations in the gaseous phase of a workshop ranging from 32 to 4505 ng m^{-3} [239]. The low affinity of these compounds towards the particulate matter is highlighted since the percentage of musks in the gas phase to the total was higher than 97 %. Synthethic musks have been recently determined in indoor air samples from homes of North-western Spain by Regueiro *et al* [50]. Measured concentrations of HHCB and AHTN were from 143 to 1129 ng m^{-3}, and from

21 to 77 ng m^{-3}, respectively. Celestolide (4-acetyl-1,1-dimethyl-6-tert. − butylindane, ADBI) and AHMI were also found in one sample at concentrations of 2.6 and 8.5 ng m^{-3}, respectively, while Cashmeran (6,7-dihydro-1,1,2,3,3-pentamethyl-4(5H)indanone, DPMI), Traseolide (5-acetyl-1,1,2,6-tetramethyl-3-iso- propyldihydroindane, ATII), and musk moskene (4,6-dinitro-1,1,3,3,5-pentamethylindane, MM), were not found in any of the samples. These concentrations were higher than those measured by Kallenborn *et al* [155], but in the same order of those reported by Fromme *et al* [77] in the air of German kindergartens.

6.4. PESTICIDES

Studies focussing on the assessment of pesticide exposure and on the adverse effects of pesticides on both human health and the environment are a matter of a growing scientific and public concern [241, 242]. Pesticides are broadly used in farming for their economic benefits to fight crop pests and reduce competition from weeds, thus improving yields and protecting the quality, reliability and price of production. The widespread use of these compounds has resulted in contamination of environmental compartments, such as surface water, groundwater, soil, and air [243-246]. Interest and demand for ambient air analysis has increased the number and diversity of pesticides of concern.

For the past three decades, organophosphorus pesticides (OPPs) have been the insecticides most commonly used both by professional pest control bodies and homeowners [247]. Nevertheless, the decision of the US Environmental Protection Agency (EPA) to phase out certain uses of the organophosphate insecticides because of their potentially toxic effects to humans has led to their gradual replacement by other pesticides such as pyrethroids. Synthetic pyrethroids have been manufactured since 1950´s based upon the structure of natural pyrethrins, which are chemicals with active insecticidal properties in the pyrethrum extract (a mixture of chemicals found in certain chrysanthemum flowers). Pyrethrins are very unstable in the environment, due to oxidation and UV-radiation.

Pyrethroids are widely applied as insecticides in households and greenhouses as well as for the protection of crops. Their advantageous properties like their non-persistence in the environment (most are degraded rapidly by sunlight or other compounds found in the atmosphere), excellent insecticidal activity against a wide range of insect pests, low dose of application needed and relatively low mammalian toxicity, have dramatically increased their use in recent years [248, 249]. Therefore, they are effective substitutes of some traditional insecticides (e.g. organochlorine or organophosphate compounds) whose use is strongly or partially restricted by countries regulations. Nevertheless, laboratory tests showed that pyrethroids are toxic for fish, aquatic arthropods and honey-bees [250, 251].

Releases to the air represent the most important emission pathway for pyrethroids. Because of that, inhalation is an important route of exposure for humans, especially after spraying application in indoor air or for people living in agricultural areas in ambient air. OSHA established the occupational exposure limit for an 8-hour workday, 40-hour workweek at 5 mg of pyrethrins and pyrethroids per cubic meter of workplace air (5 mg/m^3).

Pyrethroids available today include, for example, allethrin, phenothrin, tetramethrin and cyphenothrin usually for household insects or cypermethrin, deltamethrin, permethrin, λ-cyhalothrin and cyfluthrin mainly for agricultural insects. Pyrethroids are often commercially combined with synergist compounds like piperonyl butoxide, which enhance their insecticidal activity or mixed with fungicides as 2-phenylphenol or other pesticides like propoxur (a carbamate pesticide).

Table 11 summarizes recent publications where pesticides have been determined in indoor or workplace air.

Sampling and Analysis

Many types of sorbents have been described to sample pesticides from indoor or workplace air: PUF [68, 69, 72, 74, 103, 170, 254, 258], XAD-2 resin [103, 256, 257, 259], mixtures of both adsorbents [96, 103, 260], Tenax [44, 55-58, 261], Florisil [53, 145, 146], Supelpak [105], Empore disks [137], C$_{18}$ [105], or silica gel [165, 262]. NIOSH methods 5600 and 5601 collect organophosphorus and organonitrogen pesticides, respectively in OVS-2 tubes [13]. These tubes contain a quartz filter and XAD-2 resin (270 mg /140 mg). Dobson *et al* compared the efficiencies of PUF, XAD-2, XAD-4, and two different sandwich combinations; PUF/XAD-2/PUF and PUF/XAD-4/PUF for trapping currently used pesticides in the gaseous phase using high volume (hi-vol) samplers [264].

Table 11. Analytical procedures for the determination of pesticides in indoor air

Analyte	Sampling	Desorption and sample treatment	Determination	Recovery (%)	LOD	Year	Ref
3 pesticides: chlorpyrifos, malathion and methomyl	XAD-4 (1 m^3 min^{-1}, 3 h, 180 m^3)	Extraction by shaking with ethyl acetate and filtration. HPLC fractionation. Concentration, centrifugation with 5 mL ethyl acetate and final concentration to 1 mL	GC/FID, GC/ECD	72-81	0.3-1 ng m^{-3}	1990	252
6 pyrethrins and 7 pyrethroids	Silica gel (0.5 m^3 h)	Extraction with 50 mL hexane. Transfer into glass columns and elution with 50 mL hexane/ethyl acetate (1:1). Final concentration	HRGC/ECD	NR	NR	1991	165
23 pesticides	Tenax (4h, 1m^3)	Extraction by shaking with 5 mL acetone (30 min). Filtration through glass wool. Addition of acetone, concentration to 200 µL, addition of IS and final concentration to 40 µL	GC/CI-MS	50.7-110.9	0.5-30 ng m^{-3} (1 m^3 air)	1993	57
Insecticides (chlordanes)	Tenax (1-2 L min^{-1}, 50-100 L)	Thermal desorption	GC/EI-MS (SIM)	NR	0.25 ng m^{-3} (20 L)	1994	58
aldrin, dieldrin, 4 chlordanes, pentachloroanisole and HCHs	PUF (30 m^3 h^{-1}, 50-100 m^3)	Spiking with isotopically labelled IS. Soxhlet with hexane in acetone (50%). Reduction to 0.1 mL, purification using silica and final elution using 10 mL hexane, 10 mL 50% hexane in dichloromethane and 10 mL dichloromethane	GC/MS	NR	NR	1996	74

Analyte	Sampling	Desorption and sample treatment	Determination	Recovery (%)	LOD	Year	Ref
7 pesticides: 1 carbamate, 3 pyrethroids, 1 phenylsulfamide, etc	Tenax (0.528-1.261 L min^{-1}, 60 min)	Extraction by shaking with 2 mL acetone (5 min). Filtration through a paper filter, rinsing with 2 mL acetone, evaporation, and redissolution with n-hexane or acetone	GC/ECD, GC/NPD	75-89	LOQs=0.1-0.2 µg m^{-3}	1996	56
Insecticides: pyrethrins, pyrethroids and a synergist	PUF (3 m^3 h^{-1}, 10 m^3)	Ultrasound extraction with 150 mL ethyl acetate. The extracts were combined, filtered through silanized glass wool and reduced to 1 mL	GC/ECD, GC/FID, HPLC/UV	75.5-113.9	NR	1997	73
5 insecticides pyrethroids	100 mL air dissolved in 25 mL acetone using a syringe	The syringe is washed with acetone four times. The washings are combined and concentrated	GC/ECD	NR	NR	2001	253
Insecticide and acaricide (malathion and some of its metabolites)	PUF (2 L min^{-1})	Soxhlet with 100 mL acetone. Evaporation up to almost dryness. Addition of IS and dilution to 4 mL	GC/MS-MS	93.2-94.1	0.01-0.07 ng L^{-1}	2001	72
11 pesticides: 2 fungicides, 1 carbamate, 2 pyrethroids, 1 dinitroaniline, etc	Tenax (2.1 L min^{-1}, 8 h)	Incubation with 5 mL methanol (5 min) with occasional shaking and Ultrasound exttraction. After sedimentation, 1 mL of the supernatant is filtered. After addition of IS, the final volume was adjusted with water	HPLC/UV (DAD)	70-100	1.0-9.1 µg m^{-3} (1 m^3)	2001	55
39 pesticides	XAD-2 and PUF (4-9 L min^{-1}, 24 h, 4-14 m^3)	Soxhlet with 150 mL diethyl ether in hexane (6%). Addition of a deuterated surrogate, drying with sodium sulphate, concentration, and volume adjusting to 2 mL using 10% diethyl eter in hexane	GC/EI-MS (SIM)	60-150	1-6 ng m^{-3}	2003	96

Table 11. (Continued).

Analyte	Sampling	Desorption and sample treatment	Determination	Recovery (%)	LOD	Year	Ref
Chlorpyrifos	XAD-2 and PUF (3.8 L min^{-1}, 24 h)	Soxhlet with 5% diethyl eter in hexane (PUF). Extraction with 5% diethyl eter in hexane followed by shaking (XAD-2). Addition of decachlorobiphenyl as surrogate and 2,4,5-tribromobiphenyl as IS	GC/ECD, GC/MS (confirmation)	98-120	NR	2003	103
17 insecticides and acaricides: pyrethroids, organophosphates, carbamates, etc	PUF (50 L min^{-1})	Extraction soaking in 50 mL ethyl acetate, and squeezed periodically in an ultrasound bath for 2 min (PUF plugs). Concentration to 0.5 mL, filtration through a pipette with silanized glass wool, washing with 0.4 mL ethyl acetate and adjusting to 1 mL.	GC/EI-MS (SIM)	85-109 (matrix-matched calibration)	0.1-5 ng m^{-3} (10 m^3)	2003	170
29 pesticides: 9 organophosphates, 6 carbamates, 2 pyrethroids, 6 herbicides, 5 fungicides, 1 repellant	PUF (4 L min^{-1}, 48 h, 11.5 m^3) previously spiked with terphenyls-d$_{14}$ as a recovery surrogate	Soxhlet with 6% diethyl eter in hexane. Concentration to 1 mL	GC/MS	NR	0.2-0.7 ng m^{-3} (air)	2003	254
Chlorpyrifos	Supelpak	Extraction by shaking with 3 mL toluene. Concentration	GC/MS (SIM)	NR	NR	2003	255
19 insecticides, 1 synergist, 1 fungicide	Empore disk (5 L min^{-1}, 24 h, 7.2 m^3)	Ultrasound extraction with 8 mL acetone followed by shaking. Centrifugation, addition of IS and concentration to 0.3 mL	GC/EI-MS (SIM)	>85	0.1-2.0 ng m^{-3} (7.2 m^3)	2004	137

Analyte	Sampling	Desorption and sample treatment	Determination	Recovery (%)	LOD	Year	Ref
3 chlordanes and 2 nonachlors	PUF (10 L min^{-1}, 29 m^3)	Extration with 40 mL of hot (50°C) hexane-dichloromethane (4:1). Rinsing with hot hexane/dichloromethane (4:1). Concentration. Clean-up on microcolumns of silicic acid. Elution with 2 mL hexane/dichloromethane (9:1). Reduction to ~0.1 mL, addition of deuterated PAHs as IS and final reduction to ~0.05 mL	GC/EI-MS (SIM)	82-91	0.018-0.140 ng (gas phase)	2004	68
11 pesticides: pyrethroids, carbamates, organophosphorous, etc	The output knob of a 250-mL glass flask is connected to a pump and the input knob is just open to the air. A SPME fiber (PDMS) is inserted into the sampling flask through a septum and exposed to the air stream (dynamic mode, 40 min)	Thermal desorption	GC/EI-MS (ITD, SIM)	NR	0.03-76.7 µg m^{-3}	2004	183
8 pesticides: malathion, chlorpyrifos, diazinon, etc	PUF (30-40 L min^{-1}, 24 h)	Extraction with 2 mL toluene/acetone (9:1) (NIOSH 5600)	GC/MS (SIM)	NR	0.001-0.002 µg m^{-3}	2004	69

Table 11. (Continued).

Analyte	Sampling	Desorption and sample treatment	Determination	Recovery (%)	LOD	Year	Ref
38 pesticides: herbicides, pyrethroids, organophosphate and organochlorine insecticides, fungicides	PUF (24 h, 5 L min^{-1}, 7.1 m^3)	Soxhlet with 150 mL dichloromethane. Concentration to 100 μL followed by dilution in 2 mL acetone	GC/ECD, GC/TSD, HPLC/UV (DAD)	73.1-120.2	LOQs = 0.1-562 ng m^{-3}	2006	67
Insecticides (pyrethroids)	SPMDs suspended about 2 m height from floor (48 h)	Microwave-assisted extraction with hexane/acetone (1:1). Concentration, reconstitution in 5 mL hexane, and extraction with acetonitrile. Clean-up with alumina-C$_{18}$ and elution with 10 mL acetonitrile. Evaporation to dryness. Addition of IS in isooctane	GC/EI-MS-MS	61-103 (after 2nd extraction)	0.3-0.9 ng per SPMD	2006	139
4 fungicides, 1 insecticide and 1 acaricide	Supelpak or C18	Extraction by shaking with acetone or ethyl acetate. Centrifugation	GC/NPD, HPLC/UV	79-102 (Supelpak), 84-106 (C$_{18}$)	LOQ=0.2-20 μg m^{-3} (60 L)	2006	105
11 pyrethroids, 1 synergist, 1 fungicide, 1 carbamate	Tenax (100 L min^{-1}, 1 m^3)	Ultrasound extraction with 1 mL ethyl acetate	GC/MS (ITD), GC/μECD	81-114	0.03-4.1 ng m^{-3} (μECD), 1.4-9.1 ng m^{-3} (MS)	2006	44

Analyte	Sampling	Desorption and sample treatment	Determination	Recovery (%)	LOD	Year	Ref
10 pyrethroids, 1 synergist, 1 fungicide, 1 carbamate	Florisil (100 L min^{-1}, 1 m^3)	Addition of 100 μL acetone followed by HS-SPME (PA fiber, 30 min, 100°C)	GC/MS (ITD), GC/μECD	76-119	0.001-2.1 ng m^{-3} (μECD), 0.046-7.1 ng m^{-3} (MS)	2006	53
Pentachlorophenol, bisphenol-A and nonylphenol	XAD-2 (48 h, 4 L min^{-1})	Soxhlet with dichloromethane. Concentration, SPE with florisil and concentration. Addition of surrogates	GC/MS	55-120	0.09 ng m^{-3}	2007	256
Disinfectants: Quaternary Ammonium Compounds (QACs)	XAD-2 (1 Lmin^{-1}, 100 L)	Ultrasound extraction with 5 mL acetonitrile	IC (Cationic preconcentration on column), LC/MS-MS	99.83-101.00	28 μg m^{-3} (100 L, IC), 5 ng m^{-3} (100 L, LC/MS-MS)	2007	257

The sandwiches were only slightly more efficient than XAD-2 and XAD-4 resins, followed by PUFs, and taking into account that losses of pumping efficiency were found using the sandwich designs, XAD-2 was the adsorbent recommended. Tsiropoulos *et al* investigated the trapping efficiency of XAD-2, XAD-4, Supelpak-2, Florisil and C_{18} for five pesticides [105]. Supelpak-2 and C_{18} were selected as the best adsorbents, based on their performance characteristics, such as sufficient trapping efficiency, no dependence on the relative humidity, extended range of concentration levels, good recoveries and storage stability. As it was previously mentioned in section 5 of this book, the use of quartz filters and Empore disks was tried to determine 92 SVOCs, including insecticides, synergists and fungicides [137]. Among them, 20 pesticides including fenthion, piperonyl butoxide, allethrin or tetramethrin, could not be sufficiently collected by the disks due to low retention efficiencies; while other pesticides, such as fenitrothion, pentachlorophenol or deltamethrin, could not be accurately quantified since their calibration curves were not linear. Elflein *et al* also underlined recovery problems when sampling household insecticides by means of a glass filter and two polyurethane foam cartridges [170], assuming a decomposition mechanism on the filter during the spiking experiment for four pyrethroids. In addition, these authors sentenced that polyurethane foam contributes to the "matrix-induced chromatographic response enhancement".

As it was previously described in this book, when adsorbents are used for sampling, calibration is usually performed by direct spiking of adsorbents with solutions of known concentrations of the target analytes. Another procedure to calibrate, in this case, trifluralin and triallate, was introduced by Cessna and Kerr [147]. A polytetrafluoroethylene (PTFE) U-tube is fortified with a solution of pesticides in hexane and immersed in a water bath at 50 °C. Then, air is continuously drawn through an U-tube at 0.1 L min^{-1} towards two mini-tubes packed with Tenax TA arranged in series. In this way, an easy and realistic calibration was feasible, simulating different concentrations of air samples.

Due to their possible workplace implications, some papers regarding outdoor air analysis should be pointed out. Cartridges with Florisil were used to estimate the leaf-air transfer of pesticides in vegetables [145, 146] or to measure atrazine and alachlor concentrations in agricultural areas [263]. Egea Gonzalez and co-workers developed a screening method to analyze more than 70 pesticides in air of urban locations surrounded by greenhouses [265]. These authors tested three different adsorbents (Tenax TA, Chromosorb 106 and Supelpak) obtaining the poorest recoveries with Supelpak.

Several authors have reviewed the passive sampling of pesticides, among other organic pollutants, in ambient air [266-268]. However, the number of passive sampling studies for collecting pesticides in indoor air is scarce. Esteve-Turrillas and co-workers sampled pyrethroid insecticides with SPMDs [139]. The membranes were suspended about 2 m height for a total time of 48 h in a dark and closed room treated with different insecticide sprays. Dai *et al* sampled chlorpyrifos (a termiticide) for one month in indoor air in houses using a passive sampler consisting of a porous PTFE tube filled with 0.75 g of Supelpak adsorbent resins [255]. Ramesh and Vijayalakshmi collected three pyrethroids in air of rooms treated with insecticides using an airtight syringe and then dissolved them in acetone [253].

The exposition of a SPME fiber to the contaminated atmosphere constitutes an alternative for sampling pesticides in indoor air. In this way, Ferrari *et al* published a multiresidue method using SPME for the determination of 11 pesticides in confined atmospheres [183]. Compounds belonged to different chemical families with a large range of saturated vapour pressures. A PDMS fiber was immersed for 40 min in a 250-mL flask through which air samples were dynamically pumped from the analysed atmosphere. As a field application, the proposed method was applied for the determination of procymidone concentrations as a function of time in a greenhouse. The use of SPME fibres completely avoids the use of solvents and can be applied to determine pesticide concentrations in workplace environments, like in the breathing zone of workers in greenhouses.

Paschke *et al* compared the applicability of SPME and SPMDs for semi-volatile chlorinated organic compounds in a landfill, where large amounts of lindane by-products were deposited in the past, together with other hazardous chemical residues [269]. Both samplers yielded to comparable TWA air concentrations of lindane and its isomers and of dichloro-diphenyl-trichloroethane with its metabolites. Cisper and Hemberger developed a method for the on-line detection of SVOCs, including pesticides, using membrane introduction mass spectrometry (MIMS) [131], clearly expanding the practical limits of MIMS analysis. The method used a composite membrane made by plasma deposition of a thin PDMS layer on a microporous polypropylene support fiber. Air sample flowed over the outside of the fibres counter current to the helium flow. Concentrations were found in the pptv range.

When analytes were retained on an adsorbent, an appropriate solvent is required, usually at high volumes, to quantitatively elute them. In addition to the time-consuming steps for concentration and clean up of the organic extracts, including the risk of analyte losses, the possible photodecomposition of some pesticides has been reported [137, 165, 170], showing that the determination of certain pesticides in air might require performing a rapid and careful trapping-extraction process. Classical extraction processes such as Soxhlet has extensively been used [67, 72, 74,96, 103, 254, 256], as well as solvent extraction with acetone [56, 57, 105], methanol [263, 270], acetonitrile [259], ethyl acetate [105, 252], hexane and dichloromethane [68], toluene [255], or mixtures of solvents [69], usually followed by shaking for several minutes. As a consequence of the large volumes of solvents used in this kind of extractions, a further concentration step may be required, as well as drying with anhydrous sodium sulphate [145]. Filtration through silanized glass wool [57, 73], HPLC fractionation [252], or cleaning procedures are also generally needed [68, 74, 139, 145, 256, 260].

Besides the conventional solvent extraction procedures, other techniques have been proposed for the extraction of pesticides from the trapping sorbents. Extraction of analytes is sometimes helped by sonication, usually over a period of no more than 15 minutes [44, 73, 137, 146, 170, 257, 258, 265]. Reduction in the amount of trapping sorbent would allow the reduction in solvent volumes. In this way, a recent paper describes the use of only 25 mg Tenax as trapping sorbent as part of a method to determine several pesticides in indoor air based on US-assisted extraction with a volume of ethyl acetate as low as 1 mL [44]. Detection limits for this simple and fast method ranged from 0.03 to 4.1 ng m^{-3} (1 m^3), with no need of concentration or further treatment of the extracts. Another approach based on the use of SPMDs was described for the determination of insecticides in air, requiring solvent re-extractions with 30 mL of a mixture of hexane-acetone and microwave extraction for 20 min [139]. In this procedure, concentration, reconstitution of the extracts, and different clean up steps derived from the matrix effect of SPMDs, were needed to achieve good recoveries, whereas detection limits ranged from 0.3 to 0.9 ng per membrane.

Thermal desorption is an alternative if a thermally desorbable adsorbent has been used for trapping pesticides. Some authors extracted chlordanes [58], two herbicides (trifluralin and triallate) [147], or 10 pesticides including triazines, carbamates and organochlorides, from Tenax by thermal desorption [271]. Baroja *et al* determined fenothrion and its main metabolites in forestry air by sampling on Tenax and using a thermal desorption cold trap [272].

The use of headspace (HS)-SPME has been proposed as an alternative to solvent and direct thermal desorption of pesticides, enhancing the selectivity and the sensitivity of the analysis. In such a way, Barro *et al* optimized a method for determining several pesticides in indoor air after their retention on 25 mg of Florisil [53]. After the addition of 100 µL acetone to the sorbent, a SPME was carried out by exposing a polyacrilate fiber to the HS of the vial. The fiber was then thermally desorbed in the injection port of a gas chromatograph. Using µECD detection, method detection limits as low as 0.001 ng m^{-3} were achieved for most insecticides.

The techniques of choice for the determination of pesticides in air are GC/ECD [44, 53, 56, 73, 103, 145-147, 252, 253, 260] and GC/MS [57, 58, 69, 74, 103, 170, 254, 256, 258, 260, 263, 272], although other less common detectors such as thermoionic specific detector (TSD) [67], or NPD [56, 105] have also been used with GC. When higher sensitivity is required, GC/MS-MS [72] can also be used. Egea Gonzalez *et al* determined 70 pesticides by GC/MS-MS using a large volume injection technique [265]. Injecting a higher volume of sample extract (10 µL) the sensitivity was enhanced, achieving LOQ values ranging from 0.2 for chlorothalonil to 27 ng m^{-3} for cypermethrin, based on a 1.44 m^3 air sampled.

In addition, the use of HPLC/UV [67, 73, 105, 259] has also been reported. Vincent *et al* determined quaternary ammonium biocide compounds by ion chromatography (IC) and LC/MS-MS [257]. In this particular case, IC appears to be a good alternative since it is not expensive and its use is very simple compared to LC/MS-MS. Moreover, the limit of detection could be reduced by a factor of 100 with an injection volume of 50 µL.

Concentration in Indoor Air

Pesticide control indoors is getting increasing attention. Concentrations of common household pesticides are generally higher indoors than outdoors [273]. Class and Kintrup determined household insecticides in commercial formulations, residues, surfaces, and in air during and after indoor application [165]. The concentrations of insecticides in air and their deposits on surfaces (up to 1000 µg m^{-3}) revealed possible exposure of humans by inhalation or skin adsorption. Electrically heated evaporators cause allethrin concentrations in air of 2-5 µg m^{-3} during application; much higher concentrations (300 µg m^{-3} and more) were observed when pyrethroids and other insecticides were sprayed as aerosols into a room. The insecticides laid on surfaces and some readily formed transformation products persisted for 60 h or longer. Berger-Preiss *et al* monitored the concentrations of two pyrethroids, pyrethrum and the synergist piperonyl butoxide in a model house over a period of two years after simulated pest control against cockroaches [73]. Only the pyrethrins decreased rapidly, mainly by photodecomposition. Deltamethrin and permethrin levels in the gas phase were 1.5 and 8 ng m^{-3} respectively, when a normal dose was applied. Roinestad et al identified 34 pesticides in household air ranging from 5.7 to 254.7 ng m^{-3} [57]. Comparison of dichlorvos, o-phenylphenol and propoxur levels in a home were also carried out immediately after spraying (354.7, 63.0 and 434.3 ng m^{-3} respectively), and 8 weeks after application (not detected, 35.8 and 5.8 ng m^{-3}). In other study, concentrations of aldrin, dieldrin, four chlordanes, pentachloroanisole and hexachlorocyclohexanes were measured in the living area of a home and outdoors [74]. All compounds except the hexachlorocyclohexanes showed higher indoor than outdoor air concentrations, implying that their sources were in the home. Ramesh and Vijayalakshmi deployed two different mosquito coils, an aerosol sample, and two different mosquito mats containing pyrethroids in a close room [253]. Air samples were collected at different intervals ranging from 15 min to 8 h from three different positions in the room (top, middle and bottom). The concentrations of pyrethroids were initially high at the top of the room, followed by a steady decline on moving towards the floor. At the end of a 6 h period, most of the residues were below 0.1 ppb. Rudel and co-workers determined pesticides, among other EDCs in 120 homes [96]. The 90th percentile concentrations for pesticides ranged from 10 to 19 ng m^{-3} in air. The indoor prevalence of pesticides that have been banned or restricted for many years, such as DDT, chlordane, heptachlor, methoxychlor, dieldrin and pentachlorophenol, suggested that indoor degradation is negligible. Whyatt et al measured 8 pesticides in 48-h breathed out air samples collected from more than 200 mothers during pregnancy [254]. A significant correlation was observed between the levels of chlorpyrifos, diazinon and propoxur in the breathed out air and the levels of these insecticides or their metabolites in plasma samples (maternal and/or cord). The fungicide o-phenylphenol was also detected in all air samples, but it was not measured in plasma. Other studies measured pesticides in indoor air of homes, i.g. chlordanes [68], chlorpyrifos [103, 255], phenols [256], or organophosphorus pesticides [69]. Moreover, biocides as DDT, lindane, methoxychlor, among others were identified in different locations of museums [261].

The inhalational exposure to pesticides in greenhouses is considered more critical than outdoors, because greenhouse walls restrict their rapid distribution and dilution via airflow [56]. Cruz Márquez et al developed a method for assessing both likelihood and exposure of farmers to spray applications of malathion in greenhouses [72]. The malathion concentration in the breathing area during the application was found between 69.4-85.9 µg m^{-3}. Insecticides and fungicides were monitored in greenhouses for 3-4 days after application of plant protection products by manual sprayers on different types of crops (flowers and vegetables) [56]. The maximum concentration found was 28 µg m^{-3} for parathion, and after a dissipation period of several hours, the levels were greatly influenced by ventilation and temperature. The objective of Bouvier *et al* [67] was to assess the residential pesticide exposure of non-occupationally exposed adults, and to compare it with occupational exposure of subjects working indoors. The study involved 20 exposed persons, 38 insecticides, and the sampling of 19 residences, two greenhouses, three florist shops and three veterinary departments. Indoor air concentrations were often low, but could reach 200-300 ng m^{-3} for atrazine and propoxur in residences. As expected, gardeners were exposed to pesticides sprayed in greenhouses, although florists and veterinary workers were also indirectly exposed due to the former pest control operations. Pesticide measurements were up to 220 ng m^{-3} for methidathion in greenhouses, 28.6 ng m^{-3} for lindane in florist shops, and 52.9 ng m^{-3} for diazinon in veterinary departments. Other authors monitored the concentrations of widely used plant protecting agents during and after application, as well as their spatial and temporal distribution in agricultural areas [55, 260, 271, 274, 275].

6.5. PERFLUORINATED ALKYL COMPOUNDS

Perfluorinated alkyl compounds (PFAs) are a group of organic chemicals used in a variety of consumer products for water and oil resistance, including surface treatments for fabric, upholstery, carpet, paper, and leather, in fire-fighting foams, and as insecticides [276]. Many of them combine bioaccumulative potential, toxic effects and extreme persistence; thus, they are considered as candidates for the Stockholm Convention list of persistent organic pollutants (POPs) and are regarded as a new and emerging class of environmental contaminants. Perfluorooctane sulfonate (PFOS), perfluorooctanoate (PFOA) and related compounds such as perfluoralkyl sulfonamides (PFASs) and fluorotelomer alcohols (FTOHs) figure among the most widespread PFAs [24, 277]. Up to now, there are only very few available data on indoor air concentrations of PFCs, but concentrations of volatile polyfluorinated compounds appear to be considerably higher in indoor than in outdoor air.

Perfluoroalkyl sulfonamides have been collected in indoor air by both active and passive procedures (see Table 12). Active sampling has been carried out using SPE cartridges [277] or a glass filter followed by polyurethane foam plugs [84], and air volumes between 20 and 200 m^3. These compounds have also been collected by means of polyurethane foam disk passive air samplers [24]. Very recently, Shoeib *et al* [278] have developed a novel type of polyurethane foam disk impregnated with XAD-4 powder, which provides a higher sorptive capacity for organic and polar chemicals, such as the FTOHs and PFASs. Uptake rates for this sorbent-impregnated polyurethane foam disks from 1.4 to 4.6 m^3 day^{-1} were estimated for the studied compounds.

Extraction of fluorinated compounds has been mainly performed by Soxhlet [24, 84, 278] with no further clean-up after volume concentration. Analysis is usually carried out by GC/MS operated in the EI mode with SIM [24, 84] or in the PCI mode [277, 278]. Separation of PFASs can be performed with common stationary phases 5 % phenyl substituted methylpolysiloxane [24, 84], although more polar capillary columns are required for FTOHs [277, 278]. Shoeib *et al* [24] determined PFAS in indoor air with recoveries ranging from 64 to 89 %, relative standard deviation (RSD) values lower than 8 %, and LODs between 0.01 and 7.1 pg m^{-3}.

Shoeib *et al* [84] determined concentrations of PFAS in indoor air from homes and laboratories. N-methyl perfluorooctane sulfonamidoethanol (MeFOSE), widely used as a stain repellent on carpets, was the most abundant in both indoor and outdoor air, followed by N-ethyl perfluorooctane sulfonamidoethanol (EtFOSE) (see Table 2). Mean indoor concentrations of MeFOSE and EtFOSE were 2589 and 772 pg m^{-3}, respectively. These concentrations were approximately 100 times higher than their outdoor values. PFAs in indoor air from office were evaluated by Jahnke *et al* [277], obtaining values for MeFOSE and EtFOSE of 727 and 305 pg m^{-3}, respectively.

6.6. ENVIRONMENTAL TOBACCO SMOKE

Environmental tobacco smoke (ETS) contains thousands of compounds, many of which are demonstrated carcinogens [279], such as benzene and 1, 3-butadiene. In addition, other ingredients of ETS classified as possible carcinogens according to IARC are naphthalene, styrene, ethyl benzene, and isoprene. The risk of cancer can be higher in child since they are most affected by household exposure [280]. ETS also contains pyridine, toluene, limonene, phenol, and other chemicals that are considered possible harmful for humans.

Table 12. Analytical procedures for the determination of perfluorinated alkyl compounds in indoor air

Analytes	Sampling	Desorption and sample treatment	Determination	Recovery (%)	LOD	Year	Ref.
MeFOSE, EtFOSE, EtFOSA, MeFOSEA	PUF disks (21 days, uptake rate 2.5 m^3 day^{-1}: 52.5 m^3)	Soxhlet with petroleum eter, concentration to 0.5 mL, and addition of IS (Mirex)	GC/EI-MS (SIM)	64-89	0.01-7.1 pg m^{-3}	2005	24
FTOHs, MeFOSA, EtFOSA, MeFOSE, EtFOSE	Isolute ENV+ (20-100 m^3, 1.1 m^3 h^{-1})	Elution with 34 mL ethyl acetate, concentration, cahange to isooctane, concentration to 0.2 mL, and addition of IS (TCN)	GC/PCI-MS (SIM)	17-400	3-300 pg m^{-3}	2007	277
MeFOSE, EtFOSE, MeFOSEA	PUF (100-200 m^3, 400 L min^{-1})	Soxhlet with petroleum ether/acetone (1:1). Concentration to 1 mL, solvent exchange to ethyl acetate, and addition of IS (Mirex)	GC/EI-MS (SIM)	47-60	0.3-20 pg m^{-3}	2004	84
FTOHs, MeFOSE, MeFOSA, EtFOSE, EtFOSA, MeFOSEA	PUF disk impregnated with XAD-4 powder (83 days, uptake rate 1.4-4.6 m^3 day^{-1}, 116-382 m^3)	Soxhlet with petroleum ether/acetone (1:1), previous addition of ^{13}C-labeled and deuterated surrogates. Concentration to 0.5 mL, centrifugation and addition of IS (N,N-Me$_2$FOSA)	GC/PCI-MS (SIM)	86-126	NR	2008	278

A recent study conducted by Vainiotalo *et al* [37] focused on the measurement of 16 ETS components in Finnish restaurants. They used several samplers, of which a 3M (OVM 3500) diffusive one was for 3-ethenylpyridine and nicotine (sampling rate, 24 mL min^{-1}). These both compounds were also collected in charcoal tubes using a pump at a rate of 100 or 200 mL min^{-1}. Volatiles such as toluene, limonene, and pyridine, among others, were collected in Tenax; and 1,3-butadiene and isoprene, were retained in Carbopack X. The analysis of 3-ethenylpyridine and nicotine required desorption of the samplers with a toluene/pyridine solution, whereas the Tenax and the Carbopack X tubes were thermally desorbed. In all cases, a GC/MS analysis was performed.

In the above commented study concentrations of ETS-specific VOCs were similar to those detected in restaurants elsewhere [91, 281], whereas those of BTEX were found to be lower. The study indicated that ETS significantly increases the levels of several toxic impurities and that it is a source of 1,3-butadiene.

Chapter 7

CONCLUSION

The consumption of chemical products associated to the growing demand of higher comfort is not expected to decrease in the next years. Although controls on production, distribution, and utilization, are becoming stricter, new chemicals are entering our lives, which may turn our homes, schools, offices and workplaces into harmful microenvironments. Hence, continuous research on exposure to those chemicals and on their toxic effects on humans and environment is required.

The so-called emerging pollutants that can be found indoors cover compounds such as flame retardants (polybrominated diphenyl ethers, polybrominated biphenyls, or organophosphates), plasticizers (i.g. phthalates, organotin compounds), fragrances, perfluorinated alkyl compounds, together with other compounds, such as pesticides, biocides, insecticides, or the environmental tobacco smoke. For most of them ecotoxicological data are still scarce and therefore, it is difficult to predict their health effects. The growing demand on environmental monitoring by the society, and the appearance on stage of new chemicals have encouraged the development of new, more rapid and sensitive, easy of use, and less expensive methods for the analysis of these pollutants in indoor air.

The role of sorbent materials in these analytical developments and, as a consequence, in the application of the aforementioned methods in the indoor environment has been reported, and the available sorbents for sample collection, extraction/desorption techniques, clean-up procedures, determination systems, and method performance evaluation have been summarized and discussed.

Development of new and better sorbents for air sampling is on-going. Fullerenes-extracted soot (FES) can be used as low cost adsorbents for VOCs collection. Applicability of tire powder for VOCs collection has also been well demonstrated by a linear-partitioning model. Single-walled carbon nanotubes (SWCNTs), very useful in other scientist fields have been recently proposed as a novel adsorbent for collecting VOCs in air. Their large surface area and high adsorption and desorption efficiencies makes them suitable sorbents, particularly for compounds with low boiling points and strong volatility. New multi-layer sorbents are being developed as well, because they afford the opportunity to collect compounds of a wide volatility range combining different sorbents into a single multi-layer one, diminishing the effect of water on the analysis, and improving the reproducibility, and the adsorption-desorption efficiencies. Among desorption techniques, extractions based on ultrasounds or microwave radiations are attracting great attention.

Another possibility is the combination of solid sorbents with a whole air sampling technique, for example the collection of an air sample into Tedlar bags or canisters, followed by the adsorption of the target analytes onto single sorbent tubes, or into a multilayer adsorbent bed. To enhance the sensitivity of the analytical method, SPME may also act as a sample pre-concentrator after sampling using conventional methods. Thus, sampling could be carried out by whole, passive, or active methods, and then, the analytes collected during sampling could be extracted by exposing a SPME fiber.

SDPE has been recently applied for air monitoring, and it has become a fast alternative to conventional methods. The use of membranes for air analysis is also growing, and must be taken into account as another new possibility, e.g. MESI on-line systems, devices based on LDPE membranes, or MIMS determination systems.

REFERENCES

[1] Weschler, C. *J. Atmos. Environ.* 2009, 43, 153-169.
[2] Molhave, L; Clausen, G.; Berglund, B.; De Ceaurriz, J.; Kettrup, A.; Lindvall, T.; Maroni, M.; Pickering, A. C.; Risse, U.; Rothweiler, H.; Seifert, B.; Younes, M. *Indoor Air.* 1997, 7, 225-240.
[3] Brown, S. K.; Sim, M. R.; Abramson, M. J.; Gray, C. N. *Indoor Air.* 1994, 4, 123-134.
[4] World Health Organization (WHO), Environmental Health Criteria 128 (Hexachlorobenzene), International Programme on Chemical Safety, 1991, Geneva.
[5] Etzel, R. A. *Int. J. Hyg. Environ. Health.* 2007, 210, 611-616.
[6] Kotzias, D. *Exp. Toxicol. Pathol.* 2005, 57, 5-7.
[7] Camel, V.; Caude, M. *J. Chromatogr.* A 1995, 710, 3-19.
[8] Weschler, C. J.; Nazaroff, W. W. *Atmos. Environ.* 2008, 42, 9018-9040.
[9] Batterman, S. ; Jia, C.; Hatzivasilis, G. *Environ. Res.* 2007, 104, 224-240.
[10] Rudel, R. A.; Perovich, L. *J. Atmos. Environ.* 2009, 43, 170-181.
[11] Destaillats, H.; Maddalena, R. L.; Singer, B. C.; Hodgson, A. T.; McKone, T. E. *Atmos. Environ.* 2008, 42, 1371-1388.
[12] U. S. Environmental Protection Agency (US EPA), Air Toxic Methods, Technology Transfer Network, Ambient Monitoring Technology Information Center, http://www.epa.gov/ttn/amtic/airtox.html (available February 2009).
[13] National Institute for Occupational Safety and Health (NIOSH), Manual of Analytical Methods (NMAM), http://www.cdc.gov/niosh/nmam/ (available February 2009).

[14] American Society for Testing and Materials (ASTM), Standards, http://www.astm.org/Standard/index.shtml (available February 2009).
[15] European Committee for Standardization (CEN), CEN/TC 264 "Air Quality", Standards, http://www.cen.eu/CENORM/Sectors/ Technical CommitteesWorkshops/CENTechnicalCommittees/Standards.asp?param =6245&title=CEN%2FTC+264 (available February 2009).
[16] Tolnai, B.; Hlavay, J.; Möller, D.; Prümke, H. J.; Becker, H.; Dostler, M. *Microchem. J.* 2000, 67, 163-169.
[17] Ouyang, G.; Pawliszyn, J. *J. Chromatogr. A* 2007, 1168, 226-235.
[18] Dettmer, K.; Engewald, W. *Chromatographia.* 2003, 57, S339-S347.
[19] Gokhale, S.; Kohajda, T.; Schlink, U. *Sci. Total Environ.* 2008, 407, 122-138.
[20] Namiesnik, J.; Zabiegala, B.; Kot-Wasik, A.; Partyka, M.; Wasik, A. *Anal. Bioanal. Chem.* 2005, 381, 279-301.
[21] Esteve-Turrillas, F. A.; Pastor, A.; de la Guardia, M. *Anal. Chim. Acta.* 2008, 626, 21-27.
[22] Hazrati, S.; Harrad, S. *Chemosphere.* 2007, 67, 448-455.
[23] Zhu, J.; Hou, Y.; Feng, Y.; Shoeib, M.; Harner, T. *Environ. Sci. Technol.* 2008, 42, 386-391.
[24] Shoeib, M.; Harner, T.; Wilford, B. H.; Jones, K. C.; Zhu, J. *Environ. Sci. Technol.* 2005, 39, 6599-6606.
[25] Mikes, I.; Cupr, P.; Trapp, S.; Klanova, J. *Environ. Pollut.* 2009, 157, 488-496.
[26] Bohlin, P.; Jones, K. C.; Tovalin, H.; Strandberg, B. *Atmos. Environ.* 2008, 42, 7234-7241.
[27] Chaemfa, C.; Barber, J. L.; Gocht, T.; Harner, T.; Holoubek, I.; Klanova, J.; Jones, K. C. *Environ. Pollut.* 2008, 156, 1290-1297.
[28] Zhu, X.; Pfister, G.; Henkelmann, B.; Kotalik, J.; Fiedler, S.; Schramm, K. W. *Chemosphere.* 2007, 68, 1623-1629.
[29] Zhu, X.; Pfister, G.; Henkelmann, B.; Kotalik, J.; Bernhöft, S.; Fiedler, S.; Schramm, K. W. *Environ. Pollut.* 2008, 156, 461-466.
[30] Meredith, M. L.; Hites, R. A. *Environ. Sci. Technol.* 1987, 21, 709-712.
[31] Hermanson, M. H.; Hites, R. A. *Environ. Sci. Technol.* 1990, 24, 666-671.
[32] Zhao, Y.; Yang, L.; Wang, Q. *Environ. Sci. Technol.* 2008, 42, 6046-6050.
[33] Orecchio, S.; Gianguzza, A.; Culotta, L. *Environ. Res.* 2008, 107, 371-379.

[34] Gong, Y.; Eom, I. Y.; Lou, D. W.; Hein, D.; Pawliszyn, J. *Anal. Chem.* 2008, 80, 7275-7282.
[35] Harper, M. *J. Chromatogr.* A 2000, 885, 129-151.
[36] Dettmer, K.; Engewald, W. *Anal. Bioanal. Chem.* 2002, 373, 490-500.
[37] Vainiotalo, S.; Väänänen, V.; Vaaranrinta, R. *Environ. Res.* 2008, 108, 280-288.
[38] Greally, B. R.; Nickless, G.; Simmonds, P. G. *J. Chromatogr.* A 2006, 1133, 49-57.
[39] Kongtip, P.; Tangprakorn, B.; Yoosook, W.; Chantanakul, S. *J. Occup. Health.* 2008, 50, 122-129.
[40] Stanetzek, I.; Giese, U.; Schuster, R. H.; Wünsch, G. *Am. Ind. Hyg. Assoc. J.* 1996, 57, 128-133.
[41] Trabue, S. L.; Scoggin, K. D.; Li, H.; Burns, R.; Xin, H. *Environ. Sci. Technol.* 2008, 42, 3745-3750.
[42] Caro, J.; Gallego, M. *Talanta.* 2008, 76, 847-853.
[43] Van Netten, C. *Sci. Total Environ.* 2009, 407, 1206-1210.
[44] Barro, R.; Garcia-Jares, C.; Llompart, M.; Bollain, M.H.; Cela, R. *J. Chromatogr.* A 2006, 1111, 1-10.
[45] Fernández-Martínez, G.; López-Mahía, P.; Muniategui-Lorenzo, S.; Prada-Rodríguez, D.; Fernández-Fernández, E. *Water Air Soil Pollut.* 2001, 129, 267-288.
[46] Lee, C. W.; Dai, Y. T.; Chien, C. H.; Hsu, D.J. *Environ. Res.* 2006, 100, 139-149.
[47] Singer, B. C.; Hodgson, A. T.; Hotchi, T.; Ming, K. Y.; Sextro, R. G.; Wood, E. E.; Brown, N. J. *Atmos. Environ.* 2007, 41, 3251-3265.
[48] Schripp, T.; Nachtwey, B.; Toelke, J.; Salthammer, T.; Uhde, E.; Wensing, M.; Bahadir, M. Anal. Bioanal. Chem. 2007, 387, 1907-1919.
[49] Parra, M. A.; Elustondo, D.; Bermejo, R.; Santamaría, J. M. *Atmos. Environ.* 2008, 42, 6647-6654.
[50] Regueiro, J.; Garcia-Jares, C.; Llompart, M.; Lamas, J. P.; Cela, R. J. *Chromatogr.* A 2009, doi: 10.1016/j.chroma.2008.09.062 (in press).
[51] Li, J.; Feng, Y. L.; Xie, C. J.; Huang, J.; Yu, J. Z.; Feng, J. L.; Sheng, G. Y.; Fu, J. M.: Wu, M. H. *Anal. Chim. Acta.* 2009, 635, 84-93.
[52] Barro, R.; Ares, S.; Garcia-Jares, C. Llompart, M.; Cela, R. *Anal. Bioanal. Chem.* 2005, 381, 255-260.
[53] Barro, R.; Ares, S.; Garcia-Jares, C. Llompart, M.; Cela, R. J. *Chromatogr.* A 2005, 1072, 99-106.
[54] Barro, R.; Garcia-Jares, C.; Llompart, M.; Cela, R. *J. Chromatogr. Sci.* 2006, 44, 430-437.

[55] Demel, J.; Buchberger, W.; Malissa Jr., H. J. *Chromatogr. A* 2001, 931, 107-117.
[56] Siebers, J.; Mattusch, P. *Chemosphere.* 1996, 33, 1597-1607.
[57] Roinestad, K. S.; Louis, J. B.; Rosen, J. D. *J. AOAC Int.* 1993, 76, 1121-1126.
[58] Haraguchi, K.; Kitamura, E.; Yamashita, T.; Kido, A. *Atmos. Environ.* 1994, 28, 1319-1325.
[59] Berezkin, V.G.; Drugov, Y.S. Gas Chromatography in Air Pollution Analysis; Elsevier, Amsterdam, 1991.
[60] Rothweiler, H.; Wager, P. A.; Schlatter, C. *Atmos. Environ.* 1991, 25 B, 231-235.
[61] Ciccioli, P.; Cecinato, A.; Brancaleoni, E.; Frattoni, M. *J. High Resolut. Chromatogr.* 1992, 15, 75-84.
[62] Helmig, D.; Greenberg, J. P. *J. Chromatogr. A* 1994, 677, 123-132.
[63] Pellizzari, E. D.; Demian, B.; Krost, K. *J. Anal. Chem.* 1984, 56, 793-798.
[64] Helmig, D. *Atmos. Environ.* 1997, 31, 3635-3651.
[65] McClenny, W. A.; Oliver, K. D.; Jacumin Jr., H. H.; Hunter Daughtrey Jr., E. *J. Environ. Monit.* 2002, 4, 695-705.
[66] Ashworth, D. J.; Zheng, W.; Yates, S. R. *Atmos. Environ.* 2008, 42, 5483-5488.
[67] Bouvier, G.; Blanchard, O.; Momas, I.; Seta, N. *Sci. Total Environ.* 2006, 366, 74-91.
[68] Offenberg, J. H.; Naumova, Y. Y.; Turpin, B. J.; Eisenreich, S. J.; Morandi, M. T.; Stock, T.; Colome, S. D.; Winer, A. M.; Spektor, D. M.; Zhang, J.; Weisel, C. P. *Environ. Sci. Technol.* 2004, 38, 2760-2768.
[69] Lu, C.; Kedan, G.; Fisker-Andersen, J.; Kissel, J. C.; Fenske, R. A. *Environ. Res.* 2004, 96, 283-289.
[70] Elflein, L.; Berger-Preiss, E.; Levsen, K.; Wünsch, G. *J. Chromatogr. A* 2003, 985, 147-157.
[71] Hoepner, A.; Garfinkel, R.; Hazi, Y.; Reyes, A.; Ramirez, J.; Cosme, Y.; Perera, P. *Environ. Health. Perspect.* 2003, 111, 749-756.
[72] Cruz Márquez, M.; Arrebola, F. J.; Egea González, F. J.; Castro Cano, M. L.; Martínez Vidal, J. L. *J. Chromatogr. A* 2001, 939, 79-89.
[73] Berger-Preiss, E.; Preiss, A.; Sielaff, K.; Raabe, M.; Ilgen, B.; Levsen, K. *Indoor Air.* 1997, 7, 248-261.
[74] Wallace, J. C.; Brzuzy, L. P.; Simonich, S. L.; Visscher, S. M.; Hites, R. A. *Environ. Sci. Technol.* 1996, 30, 2715-2718.

[75] Alegria, H. A.; Wong, F.; Jantunen, L. M.; Bidleman, T. F.; Salvador-Figueroa, M.; Gold-Bouchot, G.; Ceja-Moreno, V.; Waliszewski, S. M.; Infanzon, R. *Atmos. Environ.* 2008, 42, 8810-8818.
[76] Mari, M.; Nadal, M.; Schuhmacher, M.; Domingo, J. L. *Chemosphere.* 2008, 73, 990-998.
[77] Fromme, H.; Lahrz, T.; Piloty, M.; Gebhart, H.; Oddoy, A.; Rüden, H. *Indoor Air.* 2004, 14, 188-195.
[78] Fromme, H.; Albrecht, M.; Drexler, H.; Gruber, L.; Schlummer, M.; Parlar, H.; Körner, W.; Wanner, A.; Heitmann, D.; Roscher, E.; Bolte, G. *Int. J.Hyg. Environ. Health.* 2007, 210, 345-349.
[79] Allen, J. G.; McClean, M. D.; Stapleton, H. M.; Nelson, J. W.; Webster, T. F. *Environ. Sci. Technol.* 2007, 41, 4574-4579.
[80] Chen, L.; Mai, B.; Xu, Z.; Peng, X.; Han, J.; Ran, Y.; Sheng, G.; Fu, J. *Atmos. Environ.* 2008, 42, 78-86.
[81] Karlsson, M.; Julander, A.; Van Bavel, B.; Hardell, L. *Environ. Int.* 2007, 33, 62-69.
[82] Cahill, T. M.; Groskova, D.; Charles, M. J.; Sanborn, J. R.; Denison, M. S.; Baker, L. *Environ. Sci. Technol.* 2007, 41, 6370-6377.
[83] Quintana, J.B.; Rodil, R.; Reemtsmad, T.; García-López, M. Rodríguez, I. *TrAC Trends in Analytical Chemistry.* 2008, 27, 904-915.
[84] Shoeib, M.; Harner, T.; Ikonomou, M.; Kannan, K. *Environ. Sci. Technol.* 2004, 38, 1313-1320.
[85] U. S. Environmental Protection Agency (U. S. EPA), Compendium of Methods for the Determination of Toxic Organic Compounds in Ambient Air, Office of Research and Development, National Risk Management Research Laboratory, Center for Environmental Research Information, 2nd Ed, Cincinnati, 1999.
[86] Weng, M.; Zhu, L.; Yang, K.; Chen, S. J. *Hazard. Mater.* 2009 doi: 10.1016/j.jhazmat.2008.08.094 (in press).
[87] Kazos, E. A.; Nanos, C. G.; Stalikas, C. D.; Konidari, C. N. *Chemosphere.* 2008, 72, 1413-1419.
[88] U. S. Environmental Protection Agency (U. S. EPA), Compendium Method TO-17. Determination of Volatile Organic Compounds in Ambient Air Using Active Sampling Onto Sorbent Tubes, Compendium of Methods for the Determination of Toxic Organic Compounds in Ambient Air, Office of Research and Development, National Risk Management Research Laboratory, Center for Environmental Research Information, 2nd Ed, Cincinnati, 1999.

[89] Su, H. J.; Chao, C. J.; Chang, H. Y.; Wu, P. C. *Atmos. Environ.* 2007, 41, 1230-1236.
[90] Kuntasal, O. O.; Karman, D.; Wang, D.; Tuncel, S. G.; Tuncel, G. *J. Chromatogr. A* 2005, 1099, 43-54.
[91] Kim, Y. M.; Harrad, S.; Harrison, R. M. *Environ. Sci. Technol.* 2001, 35, 997-1004.
[92] Davis, M. E.; Blicharz, A. P.; Hart, J. E.; Laden, F.; Garshick, E.; Smith, T. J. *Environ. Sci. Technol.* 2007, 41, 7152-7158.
[93] Sapkota, A.; Williams, D. A.; Buckley, T. J. *Environ. Sci. Technol.* 2005, 39, 2936-2943.
[94] Wu, C. H.; Lin, M. N.; Feng, C. T.; Yang, K. L.; Lo, Y. S.; Lo, J. G. *J. Chromatogr. A* 2003, 996, 225-231.
[95] Baek, S. O.; Jenkins, R. A. *Atmos. Environ.* 2004, 38, 6583-6599.
[96] Rudel, R. A.; Camann, D. E.; Spengler, J. D.; Korn, L. R.; Brody, J. G. *Environ. Sci. Technol.* 2003, 37, 4543-4553.
[97] Li, A.; Schoonover, T. M.; Zou, Q.; Norlock, F.; Conroy, L. M.; Scheff, P.A.; Wadden, R. A. *Atmos. Environ.* 2005, 39, 3491-3499.
[98] Norlock, F.; Jang, J. K.; Zou, Q.; Schoonover, T. M.; Li, A. *J. Air Waste Manage Assoc.* 2002, 52, 19-22.
[99] Rudel, R. A.; Brody, J. G.; Spengler, J. D.; Vallarino, J.; Geno, P. W.; Yau, A. *J. Air Waste Manage Assoc.* 2001, 51, 499-512.
[100] Iavicoli, I. ; Chiarotti, M.; Bergamaschi, A.; Marsili, R.; Carelli, G. *J. Chromatogr. A* 2007, 1150, 226-235.
[101] Xie, Z.; Ebinghaus, R.; Temme, C.; Lohmann, R.; Caba, A.; Ruck, W. *Environ. Sci. Technol.* 2007, 41, 4555-4560.
[102] Julander, A.; Westbergb, H.; Engwall, M. ; Van Bavel, B. *Sci. Total Environ.* 2005, 350, 151-160.
[103] Stout II, D. M.; Mason, M. A. *Atmos. Environ.* 2003, 37, 5539-5549.
[104] Ribes, A.; Carrera, G.; Gallego, E.; Roca, X.; Berenguer, M. J.; Guardino, X. *J. Chromatogr. A* 2007, 1140, 44-55.
[105] Tsiropoulos, N. G.; Bakeas, E. B.; Raptis, V.; Batistatou, S. S. *Anal. Chim. Acta.* 2006, 573-574, 209-215.
[106] Herrington, J. S.; Zhang, J. J. *Atmos. Environ.* 2008, 42, 2429-2436.
[107] Bechara, J.; Borbon, A.; Jambert, C.; Perros, P. E. *Anal. Bioanal. Chem.* 2008, 392, 865-876.
[108] J. Wang, L. Tuduri, M. Mercury, M. Millet, O. Briand, M. Montury, *Environ. Pollut.* 157 (2009) 365-370.
[109] Nicolle, J. ; Desauziers, V.; Mocho, P. *J. Chromatogr. A* 2008, 1208, 10-15.

[110] Scheppers-Wercinski, S. A.; Pawliszyn, J. *Solid Phase Microextraction;* Dekker, New York (N. Y), 1999, pp 1-26.
[111] Ai, J. *Anal. Chem.* 1997, 69, 1230-1236.
[112] Gorecki, T. In Applications of Solid Phase Microextraction; Pawliszyn, J. Ed.; RSC Chromatography Monographs. The Royal Society of Chemistry, Cambridge, 1999, pp 92-108.
[113] Namiesnik, J.; Jastrzebska, A.; Zygmunt, B. *J. Chromatogr. A* 2003, 1016, 1-9.
[114] Wei, L.; Ou, Q.; Li, J.; Liang, B. *Chromatographia.* 2004, 59, 601-606.
[115] Luo, F.; Wu, Z.; Tao, P.; Cong, Y. *Anal. Chim. Acta.* 2009, 631, 62-68.
[116] Saba, A. ; Cuzzola, A.; Raffaelli, A.; Pucci, S.; Salvadori, P. *Rapid Commun. Mass Spectrom.* 2001, 15, 2404-2408.
[117] Saba, A. ; Raffaelli, A.; Pucci, S.; Salvadori, P. *Rapid Commun. Mass Spectrom.* 1999, 13, 1899-1902.
[118] Elke, K.; Jermann, E.; Begerow, J.; Dunemann, L. *J. Chromatogr. A* 1998, 826, 317-324.
[119] Chai, M.; Pawliszyn, J. *Environ. Sci. Technol.* 1995, 29, 693-701.
[120] Lestremau, F.; Andersson, F. A. T.; Desauziers, V.; Fanlo, J. L. *Anal. Chem.* 2003, 75, 2626-2632.
[121] Lee, J. H.; Hwang, S. M.; Lee, D. W.; Heo, G. S. *Bull. Korean Chem. Soc.* 2002, 23, 488-496.
[122] Chai, M.; Tang, Y. Z. *Int. J. Environ. Anal. Chem.* 1998, 72, 77-82.
[123] Kim, K. H.; Kim, D. *Microchem. J.* 2009, 91, 16-20.
[124] Ras, M. R.; Marcé, R. M.; Borrull, F. *Talanta.* 2008, 77, 774778.
[125] Lipinski, J. *Fresenius J. Anal. Chem.* 2001, 369, 57-62.
[126] Van Durme, J.; Demeestere, K.; Dewulf, J.; Ronsse, F.; Braeckman, L.; Pieters, J.; Van Langenhove, H. *J. Chromatogr. A* 2007, 1175, 145-153.
[127] Wang, L.; Lord, H.; Morehead, R.; Dorman, F.; Pawliszyn, J. *J. Agric. Food Chem.* 2002, 50, 6281-6286.
[128] Yang, M. J.; Harms, S.; Luo, Y. Z.; Pawliszyn, J. *Anal. Chem.* 1994, 66, 1339-1346.
[129] Wennrich, L.; Popp, P.; Hafner, C. *J. Environ. Monit.* 2002, 4, 371-376.
[130] Paschke, H.; Popp, P. *Chemosphere.* 2005, 58, 855-863.
[131] Cisper, M. E.; Hemberger, P. H. *Rapid Commun. Mass Spectrom.* 1997, 11, 1449-1453.
[132] Hong, J.; Maguhn, J.; Freitag, D.; Kettrup, A. *Fresenius J. Anal. Chem.* 2001, 371, 961-965.
[133] Wei, M. C.; Jen, J. F. *J. Chromatogr. A* 2003, 1012, 111-120.

[134] Wei, M. C.; Chang, W. T.; Jen, J. F. *Anal. Bioanal. Chem.* 2007, 387, 999-1005.
[135] Ingerowski, G.; Friedle, A.; Thumulla, J. *Indoor Air.* 2001, 11, 145-149.
[136] Kallenborn, R.; Gatermann, R. In The Handbook of Environmental Chemistry; Rimkus, G. G. Ed.; Vol. 3, Part X, Springer, Berlin, 2004; pp 85-104.
[137] Yoshida, T.; Matsunaga, I.; Oda, H. *J. Chromatogr.* A 2004, 1023, 255-269.
[138] Arrebola, F.J.; Martinez Vidal, J.L.; Fernández-Gutiérrez, A. *J. Chromatogr. Sci.* 2001, 39, 177-182.
[139] Esteve-Turrillas, F.A.; Pastor, A.; De la Guardia, M. *Anal. Chim. Acta.* 2006, 560, 118-127.
[140] Bjorklund, J.; Isetun, S.; Nilsson, U. *Rapid Commun. Mass Spectrom.* 2004, 18, 3079-3083.
[141] Sjödin, A.; Carlsson, H.; Thuresson, K.; Sjölin, S.; Bergman, A.; Östman, C. *Environ. Sci. Technol.* 2001, 35, 448-454.
[142] Santos, F. J.; Galceran, M. T. *TrAC Trends in Analytical Chemistry.* 2002, 21, 672-685
[143] Kierkegaard, A.; Bjorklund, J.; Friden, U. *Environ. Sci. Technol.* 2004, 38, 3247-3253.
[144] Regueiro, J.; Llompart, M.; Garcia-Jares, C.; Cela, R. *Anal. Bioanal. Chem.* 2007, 388, 1095-1107.
[145] Yang, X.; Jiang, X.; Yu, G.; Yao, F.; Bian, Y.; Wang, F. *Environ. Pollut.* 2007, 148, 555-561.
[146] Castro, J.; Pérez, R.A.; Miguel, E.; Sánchez-Brunete, C.; Tadeo, J.L. *J. Chromatogr.* A 2002, 947, 119-127.
[147] Cessna, A. J.; Kerr, L. A. *J. Chromatogr.* 1993, 642, 417-423.
[148] Polo, M.; Garcia-Jares, C. M.; Llompart, M.; Cela, R. *Anal. Bioanal. Chem.* 2007, 388, 1789-1799.
[149] Pradhan, M.; Aziz, M. S. I.; Grilli, R.; Orr-Ewing, A. J. *Environ. Sci. Technol.* 2008, 42, 7354-7359.
[150] Moliner-Martínez, Y.; Campíns-Falcó, P.; Herráez-Hernández, R. *J. Chromatogr.* A 2004, 1059, 17-24.
[151] U. S. Environmental Protection Agency (U. S. EPA), Compendium Method TO-11A. Determination of formaldehyde in Ambient Air Using Adsorbent Cartridge Followed by High Performance Liquid Chromatography (HPLC). Active Sampling Methodology, Compendium of Methods for the Determination of Toxic Organic Compounds in Ambient Air, Office of Research and Development, National Risk

Management Research Laboratory, Center for Environmental Research Information, 2nd Ed, Cincinnati, 1999.
[152] E. R. Kennedy, T. J. Fischbach, R. Song, P. M. Eller, S. A. Shulman, Guidelines for Air Sampling and Analytical Method Development and Evaluation, National Institute for Occupational Safety and Health (NIOSH), U. S. Department of Health and Human Services, Public Health Service, Center for Disease Control and Prevention, Cincinnati, 1995.
[153] McClenny, W. A.; Oliver, K. D.; Jacumin Jr., H. H.; Hunter Daughtrey Jr., E.; Whitaker, D. A. *J. Environ. Monit.* 2005, 7, 248-256.
[154] Toda, H.; Sako, K.; Yagome, Y.; Nakamura, T. *Anal. Chim. Acta.* 2004, 519, 213-218.
[155] Kallenborn, R.; Gatermann, R.; Planting, S.; Rimkus, G.G.; Lund, M.; Schlabach, M.; Burkow, I.C. *J. Chromatogr.* A 1999, 846, 295-306.
[156] Figge, K.; Rabel, W.; Wiek, A. Fresenius Z. *Anal. Chem.* 1987, 327, 261-278.
[157] Baya, M. P.; Siskos, P. A. *Analyst.* 1996, 121, 303-307.
[158] Peters, R. J. B.; Bakkeren, H. A. *Analyst.* 1994, 119, 71-74.
[159] Harper, M. *Ann. Occup. Hyg.* 1993, 37, 65-88.
[160] Sydor, R.; Pietrzyk, D. *J. Anal. Chem.* 1978, 50, 1842-1847.
[161] Pierini, E.; Sampaolo, L.; Mastrogiacomo, A. R. *J. Chromatogr.* A 1999, 855, 593-600.
[162] Prado, C.; Alcaraz, M. J.; Fuentes, A.; Garrido, J.; Periago, J. F. *J. Chromatogr.* A 2006, 1129, 82-87.
[163] Liu, J. M.; Lin, L.; Fan, H. L.; Ning, Z. W.; Zhao, P. *Chin. J. Anal. Chem.* 2007, 35, 830-838.
[164] Strömvall, A. M.; Petersson, G. *J. Chromatogr.* A 1992, 589, 385-401.
[165] Class, T. J.; Kintrup, J. Fresenius J. *Anal. Chem.* 1991, 340, 446-453.
[166] Class, T. J. Fresenius J. *Anal. Chem.* 1992, 342, 805-808.
[167] Fernandez-Alvarez, M.; Sanchez-Prado, L.; Lores, M.; Llompart, M.; Garcıa-Jares, C.; Cela, R. J. Chromatogr. A, 2007, 1152, 156-167.
[168] Zanella, R.; Schilling, M.; Klockow, D. *J. Environ. Monit.,* 1999, 1, 441–443.
[169] Pellizzari, E. D.; Krost, K. J. Anal. Chem. 1984, 56, 1813-1819.
[170] International Union of Pure and Applied Chemistry (IUPAC); The Gold Book. Compendium of Chemical Terminology, 2nd Ed., Blackwell Scientific Publications, Oxford, 1997, http://goldbook.iupac.org/ (available February 2009).

[171] Lodge jr, J. P. Methods of air sampling and analysis. Intersociety Committee, CRC Press, Taylor and Francis, London, 1988.
[172] Desauziers, V. *Trends Anal. Chem.* 2004, 23, 252-260.
[173] Barro, R.; Ares, S.; Garcia-Jares, C.; Llompart, M.; Cela, R. J. *Chromatogr. A* 2004, 1045, 189-196.
[174] Otaka, T.; Yoshinaga, J.; Yanagisawa, Y. *Environ. Sci. Technol.* 2001, 35, 3099-3102.
[175] Tumbiolo, S.; Gal, J. F.; Maria, P. C.; Zerbinati, O. *Anal. Bioanal. Chem.* 2004, 380, 824-830.
[176] Chen, Y.; Koziel, J. A.; Pawliszyn, J. *Anal. Chem.* 2003, 75, 6485-6493.
[177] Tuduri, L.; Desauziers, V.; Fanlo, J. L. *Microcol. Sep.* 2000, 12, 550-557.
[178] Bartelt, R. J.; Zilkowski, B. W. *Anal. Chem.* 1999, 71, 92-101.
[179] Bartelt, R. J.; Zilkowski, B. W. *Anal. Chem.* 2000, 72, 3949-3955.
[180] Bartelt, R. J. *Anal. Chem.* 1997, 69, 364-372.
[181] Gorlo, D.; Wolska, L.; Zygmunt, B.; Namiesnik, J. *Talanta.* 1997, 44, 1543-1550.
[182] Mangani, F.; Cenciarini, R. *Chromatographia.* 1995, 41, 678-684.
[183] Ferrari, F.; Sanusi, A.; Millet, M.; Montury, M. *Anal. Bioanal. Chem.* 2004, 379, 476-483.
[184] Barro, R.; Regueiro, J.; LLompart, M.; Garcia-Jares, C. *J. Chromatogr. A* 2009, 1216, 540-566.
[185] Garcia-Jares, C.; Regueiro, J.; Barro, R.; Dagnac, T.; LLompart, M. *J. Chromatogr. A* 2009, 1216, 567-597.
[186] Mandalakis, M.; Atsarou, V. ; Stephanou, E. G. *Environ. Pollut.* 2008, 155, 375-382.
[187] Covaci, A. ; Voorspoels, S. ; De Boer, J. *Environ. Inter.* 2003, 29, 735-756.
[188] De Wit, C. *Chemosphere.* 2002, 46, 583-624.
[189] Sjödin, A.; Päpke, O.; McGahee, E.; Jones, R.; Focant, J.F.; Pless-Mulloli, T.; Toms, L.M.; Wang, R.; Zhang, Y.; Needham, L.; Herrmann, T.; Patterson, D. *Organohal. Comp.* 2004, 66, 3770-3775.
[190] Harrad, S.; Diamond, M. *Atmos. Environ.* 2006, 40, 1187-1188.
[191] Jones-Otazo, H.A.; Clarke, J.P.; Diamond, M.L.; Archbold, J.A.; Ferguson, G.; Harner, T.; Richardson, G.M.; Ryan, J.J.; Wilford, B. *Environ. Sci. Technol.* 2005, 39, 5121-5130.
[192] Currado, G.M.; Harrad, S. *Environ. Sci. Technol.* 2000, 34, 78-82.
[193] Päpke, O.; Fürst, P.; Herrmann, T. *Talanta.* 2004, 63, 1203-1211.

[194] Covaci, A.; Voorspoels, S.; Ramos, L.; Neels, H.; Blust, R. *J. Chromatogr. A* 2007, 1153, 145-171.
[195] Thomsen, C.; Leknes, H.; Lundanes, E.; Becher, G. *J. Chromatogr. A* 2001, 923, 299-304.
[196] Herrmann, T.; Schilling, B.; Päpke, O. *Organohalogen. Compd.* 2003, 63, 361-364.
[197] De Boer, J.; Wells, D.E. *Trends Anal. Chem.* 2006, 25, 364-372.
[198] De Boer, J.; Cofino, W.P. *Chemosphere.* 2002, 46, 625-633.
[199] Bjorklund, J.; Tollback, P.; Ostman, C. *J. Sep. Sci.* 2003, 26, 1104-1110.
[200] Pettersson-Julander, A.; Van Bavel, B.; Engwall, M.; Westberg, H. *J. Environ. Monit.* 2004, 6, 874-880.
[201] Harrad, S.; Wijesekera, R.; Hunter, S.; Halliwell, C.; Baker, R. *Environ. Sci. Technol.* 2004, 38, 2345-2350.
[202] Tollbäck, J.; Creszenci, C.; Dyremark, E. *J. Chromatogr. A* 2006, 1104, 106-112.
[203] Wilford, B.H.; Harner, T.; Zhu, J.; Shoeib, M.; Jones, K.C. *Environ. Sci. Technol.* 2004, 38, 5312-5318.
[204] Harrad, S.; Hazrati, S.; Ibarra, C. *Environ. Sci. Technol.* 2006, 40, 4633-4638.
[205] Gevao, B.; Al-Bahloul, M.; Al-Ghadban, A.N.; Ali, L.; Al-Omair, A.; Helaleh, M.; Al-Matrouk, K.; Zafar, J. *Atmos. Environ.* 2006, 40, 1419-1426.
[206] Butt, C.M.; Diamond, M.L.; Truong, J.; Ikonomou, M.G.; Ter Schure, A.F.H. *Environ. Sci. Technol.* 2004, 38, 724-731.
[207] Saito, I.; Onuki, A.; Seto, H. *Indoor Air.* 2007, 17, 28-36.
[208] Björklund, J. Gas chromatography and mass spectrometry of polybrominated diphenyl ethers. Ph. D. Thesis, Department of Analytical Chemistry, Stockholm University, Stockholm, Sweden, 2003.
[209] Covaci, A. ; De Boer, J. ; Ryan, J.J. ; Voorspoels, S.; Schepens, P. *Anal. Chem.* 2002, 74, 790-798.
[210] Björklund, J.; Tollbäck, P.; Hiärne, C.; Dyremark, E.; Östman, C. *J. Chromatogr. A* 2004, 1041, 201-210.
[211] Carlsson, H. ; Nilsson, U. ; Östman, C. *Environ. Sci. Technol.* 2000, 34, 3885-3889.
[212] Hartmann, P.C.; Burgi, D.; Giger, W. *Chemosphere.* 2004, 57, 781-787.
[213] WHO, Environmental Health Criteria 112 (Tri-n-butyl phosphate) International Programme on Chemical Safety, WHO, Geneva, 1991.

[214] WHO, Environmental Health Criteria 209 (Flame retardants: tris(chloropropyl) phosphate and tris(2-chloroethyl) phosphate). International Programme on Chemical Safety, WHO, Geneva, 1998.
[215] Carlsson, H.; Nilsson, U.; Becker, G.; Östman, C. *Environ. Sci. Technol.* 1997, 31, 2931-2936.
[216] Marklund, A.; Andersson, B.; Haglund, P. *J. Environ. Monit.* 2005, 7, 814-819.
[217] Staaf, T.; Östman, C. *J. Environ. Monit.* 2005, 7, 344-348.
[218] Isetun, S.; Nilsson, U.; Colmsjö, A.; Johansson, R. *Anal. Bioanal. Chem.* 2004, 378, 1847-1853.
[219] Isetun, S.; Nilsson, U.; Colmsjö, A. *Anal. Bioanal. Chem.* 2004, 380, 319-324.
[220] Isetun, S.; Nilsson, U. *Analyst.* 2005, 130, 94-98.
[221] Sanchez, C.; Ericsson, M.; Carlsson, H.; Colmsjö, A. *J. Chromatogr. A* 2003, 993, 103-110.
[222] Frankhauser, A.; Grob, K. *Anal. Chim. Acta.* 2007, 582, 353-360.
[223] Gomez-Hens, M.; Aguilar-Ceballos, M.P. *Trends Anal. Chem.* 2003, 22, 847-857.
[224] Tienpont,B.; David, D.; Sandra, P.; Vanwalleghem, F. *J. Microcol. Sep.* 2000, 12, 194-203
[225] Llompart, M.; García-Jares, C.; Landín, P. In Chromatographic Analysis of the Environment; Nollet, L.M.L. Ed.; 3rd Ed, CRC Press, Boca Raton, FL, 2007, pp 1103-1154.
[226] Kang, Y.; Den, W.; Bai, H.; Ko, F.H. *J. Chromatogr. A* 2005, 1070, 137-145.
[227] Kawata, K.; Minagawa, M.; Fujieda, Y. *J. Chromatogr.* 1993, 653, 369-373.
[228] Xie, Z.; Selzer, J.; Ebinghaus, R.; Caba, A.; Ruck, W. *Anal. Chim. Acta.* 2006, 546, 198-207.
[229] Peijnenburg, W.J.G.M., Struijs, J. *Ecotoxicol. Environ. Safety.* 2006, 63, 204-215.
[230] Wang, X.K. *Chin. Chem. Lett.* 2002, 13, 557-560.
[231] Thuren, A.; Larsson, P. *Environ. Sci. Technol.* 1990, 24, 554-559.
[232] Helmig, D. *J. Chromatogr. A* 1999, 843, 129-146.
[233] Sheldon, L.; Clayton, A.; Keever, J.; Perritt, R.; Whitaker, D. PTEAM: Monitoring of Phthalates and PAHs in Indoor and Outdoor Air Samples in Riverside, California: Final Report, Volume II, California Air Resources Board, Research Division, Contract No. A933-144, Sacramento, CA, December 1992.

[234] Cooke, G.M. *Toxicol. Lett.* 2002, 126, 121-130.
[235] Kallenborn, R.; Gatermann, R.; Rimkus, G.G. *J. Environ. Monit.* 1999, 1, 70N-74N.
[236] US EPA, High production volume (HPV) Chemical List database June 8, 2003 (http:/www.epa.gov/chemrtk/opptsrch.htm).
[237] Peck, A.M.; Hornbuckle, K.C. *Environ. Sci. Technol.* 2004, 38, 367-372.
[238] Bridges, B. *Flavour Fragr. J.* 2002, 17, 361-371.
[239] Chen, D.; Zeng, X.; Sheng, Y.; Bi, X.; Gui, H.; Sheng, G.; Fu, J. *Chemosphere.* 2007, 66, 252-258.
[240] Buerge, I.J. ; Buser, H.R. ; Muller, M.D. ; Poiger, T. *Environ. Sci. Technol.* 2003, 37, 5636-5644.
[241] Dorr, G.; Noller, B.; Woods, N.; Hewitt, A.; Hanan, J.; Adkins, S.; Ricci, P.F. Rational Environmental Management of Agrochemicals; ACS Symposium Series, vol. 966, 2007, pp 53-65.
[242] Ritter, L.; Goushleff, N. C. I.; Arbuckle, T.; Cole, D.; Donald; D; Raizenne, M. J. *Toxicol. Env. Health, Part B: Critical Reviews.* 2006, 9, 441-456.
[243] Tariq, M.I.; Afzal, S. ; Hussain, I. *Environ. Int.* 2004, 30, 471-479.
[244] Fernandez-Alvarez, M.; Llompart, M.; Lamas, J.P.; Lores, M.; Garcia-Jares, C.; Cela, R.; Dagnac, T. *J. Chromatogr.* A 2008, 1188, 154-163.
[245] Xue, N.; Xu, X.; Jin, Z. *Chemosphere.* 2005, 61, 1594-1606.
[246] Harner, T.; Shoeib, M.; Kozma, M.; Gobas, F.A.; Li, S.M. *Environ. Sci. Technol.* 2005, 39, 724-731.
[247] Jeannot, R.; Dagnac, T. In: Chromatographic Analysis of the Environment; L.M.L. Nollet Ed.; Chromatographic Science Series, vol. 93, Taylor & Francis, Boca Raton, Florida, USA, 2006, pp 841-888.
[248] Agency for Toxic Substances and Disease Registry (ATSDR). Toxicological Profile for pyrethrins and pyrethroids, U.S. Department of Health and Human Services, Public Health Service, 2003.
[249] World Health Organization (WHO). Environmental Health Criteria Nos. 64, 82, 87, 94, 96-99 and 142, The International Programme on Chemical Safety, United Nations Environment Programme, International Labour Organisation and World Health Organization, Geneva, pp1986-1992.
[250] Clark, J. R.; Goodman, L. R.; Borthwick, P. W.; Jr. Patrick, J. M.; Cripe, G. M. ; Moody, P. M. *Environ. Toxicol. Chem.* 1989, 8, 393-401.
[251] Cotas, J. R.; Symonik, D. M.; Bradbury, S. P.; Dyer, S. D.; Timson, L. K.; Atchison, G. J. *Environ. Toxicol. Chem.* 1989, 8, 671-679.

[252] Seiber, J.N.; Glotfelty, D.E; Lucas, A.D.; McChesney, M.M.; Sagebiel, J.C.; Wehner, T.A. *Arch. Environ. Contam. Toxicol.* 1990, 19, 583-592.
[253] Ramesh, A.; Vijayalakshmi, A. *J. Environ. Monit.* 2001, 191-193.
[254] Whyatt, R.M. ; Barr, D.B. ; Carmann, D.E. ; Kinney, P.L. ; Barr, J.R. ; Andrews, H.F. ; Hoepner, L.A. ; Garfinkel, R. ; Hazi, Y. ; Reyes, A. ; Ramirez, J. ; Cosme, Y. ; Perera, P. *Environ. Health. Perspect.* 2003, 111, 749-756.
[255] Dai, H. ; Asakawa, F. ; Suna, S. ; Hirao, T. ; Karita, T. ; Fukunaga, I. ; Jitsunari, F. *Environ. Health Prev. Med.* 2003, 8, 139-145.
[256] Wilson, N.K. ; Chuang, J.C. ; Morgan, M.K. ; Lordo, R.A. ; Sheldon, L.S. *Environ. Res.* 2007, 103, 9-20.
[257] Vincent, G. ; Kopferschmitt-Kubler, M.C. ; Mirabel, P.; Pauli, G.; Millet, M. *Environ. Monit. Assess.* 2007, 133, 25-30.
[258] Berger-Preiss, E. ; Elflein, L. *Methods Biotechnol.* 2006, 19, 179-190.
[259] Kennedy, E.R. ; Lin, J.J. ; Reynolds, J.M. ; Perkins, J.B. *Am. Ind. Hyg. Assoc. J.* 1997, 58, 720-725.
[260] Yao, Y. ; Tuduri, L. ; Harner, T. ; Blanchard, P. ; Waite, D. ; Poissant, L. ; Murphy, C. ; Belzer, W. ; Aulagnier, F. ; Li, Y.F. ; Sverko, E. *Atmos. Environ.* 2006, 40, 4339-4351.
[261] Schieweck, A. ; Delius, W. ; Siwinski, N. ; Vogtenrath, W. ; Genning, C. ; Salthammer, T. *Atmos. Environ.* 2007, 41, 3266-3275.
[262] Watanabe, T. *Bull. Environ. Contam. Toxicol.* 1998, 60, 669-676.
[263] Roach, C. E.; Anderson, L. G.; Foreman, W. T. In Proceedings of the AWMA Annual Meeting, 90th, 1997, Paper No. 97-RA122.03.
[264] Dobson, R.; Scheyer, A.; Rizet, A.L.; Mirabel, P.; Millet, M. *Anal. Bioanal. Chem.* 2006, 386, 1781-1789.
[265] Egea-Gonzalez, F.J.; Mena-Granero, A.; Glass, C.R.; Garrido-Frenich, A.; Martinez Vidal, J.L. *Rapid Commun. Mass Spectrom.* 2004, 18, 537-543.
[266] Bohlin, P.; Jones, K.C. ; Strandberg, B. *J. Environ. Monit.* 2007, 9, 501-509.
[267] Esteve-Turrillas, F.A.; Yusá, V.; Pastor, A.; De la Guardia, M. *Talanta.* 2008, 74, 443-457.
[268] Partyka, M.; Zabiegala, B.; Namiesnik, J. *Crit. Rev. Anal. Chem.* 2007, 37, 51-78.
[269] Paschke, A.; Vrana, B.; Popp, P.; Schürmann, G. Environ. Pollut. 2006, 144, 414-422.
[270] Oepkemeier, S.; Schreiber, S.; Breuer, D.; Key, G.; Kleiböhmer, W. *Anal. Chim. Acta.* 1999, 393, 103-108.

[271] Briand, O.; Millet, M.; Bertrand, F. ; Clément, M.; Seux, R. *Anal. Bioanal. Chem.* 2002, 374, 848-857.
[272] Baroja, O. ; Unceta, N. ; Sampedro, M.C. ; Goicolea, M.A. ; Barrio, R.J. *J. Chromatogr.* A 2004, 1059, 165-170.
[273] Mukerjee, S.; Ellenson, W.D.; Lewis, R.G.; Stevens, R.K.; Somerville, M.C.; Shadwick, D.S.; Willis, R.D. *Environ. Int.* 1997, 23, 657-673.
[274] Baraud, L.; Tessier, D.; Aaron, J.J.; Quisefit, J.P.; Pinart, J. *Anal. Bioanal. Chem.* 2003, 377, 1148-1152.
[275] Sanusi, A.; Millet, M.; Mirabel, P.; Wortham, H. *Sci. Total Environ.* 2000, 263, 263-277.
[276] Giesy, J.P.; Kannan, K. *Environ. Sci. Technol.* 2002, 36, 146A-152A.
[277] Jahnke, A.; Huber, S.; Temme, C.; Kylin, H.; Berger, U. *J. Chromatogr.* A 2007, 1164, 1-9.
[278] Shoeib, M.; Harner, T.; Lee, S.C.; Lane, D.; Zhu, J. *Anal. Chem.* 2008, 80, 675.
[279] IARC, IARC Monographs on the Evaluation of the carcinogenic risks to humans. *Tobacco smoke and involuntary smoking,* vol. 83, Lyon, 2004, pp 1012-1070.
[280] Irigaray, P. ; Newby, J. A.; Clapp, R.; Hardell, L.; Howard, V.; Montagnier, L.; Epstein, S.; Belpomme, D. *Biomed. Pharmacotherapy.* 2007, 61, 640-658.
[281] Higgins, C. E.; Thompson, C. V.; Jigner, R. H.; Jenkins, R. A.; Guerin, M. R. Determination of vapor phase hydrocarbons and nitrogen-containing constituents in environmental tobacco smoke. Internal Progress Report. Analytical Chemistry Division, Oak Ridge National Laboratory, Oak Ridge, 1990.

GENERAL ABBREVIATIONS

BTV	breakthrough volume
ETS	environmental tobacco smoke
FES	fullerenes-extracted soot
HS	head-space
I.D.	internal diameter
I/O	indoor/outdoor ratios
IS	internal standard
K-D	Kuderma-Danish
LDPE	low-density polyethylene
LOD	limit of detection
LOQ	limit of quantification
MESCO	membrane-enclosed sorptive coating sampler
MESI	membrane extraction with solid interface
m/z	mass to charge ratio
NTD	needle trap device
PISCES	passive in situ concentration/extraction samplers
PUF	polyurethane foam
QA/QC	quality assurance / quality control
SIF	sorbent-impregnated filters
S/N	signal to noise ratio
SPMDs	semipermeable membrane devices
SWCNTs	single-walled carbon nanotubes
TWA	time-weighted average

ORGANIZATIONS

ASTM	American Society for Testing and Materials
CEN	European Committee for Standardization
IARC	International Agency for Research on Cancer
IUPAC	International Union of Pure and Applied Chemistry
NBS	National Bureau of Standards
NIOSH	National Institute for Occupational Safety and Health
OSHA	Occupational Safety and Health Administration
US EPA	United States Environmental Protection Agency

TECHNIQUES

AED	atomic electron detection/detector
APCI	atmospheric pressure ionization
ECD	electron capture detection/detector
ESI	electrospray ionization
FPD	flame photoionization detection/detector
GC	gas chromatography
GPC	gel permeation chromatography
HPLC	high-pressure liquid chromatography
IC	ion chromatography
ITD	ion trap detector
MAE	microwave assisted extraction
MIMS	membrane introduction mass spectrometry
MS	mass spectrometry
PCI	positive chemical ionization
PID	photoionization detector
PSE	pressurize solvent extraction
PTV	programmable temperature vaporizing
SE	solvent extraction
SPDE	solid-phase dynamic extraction
SPE	solid phase extraction
SPME	solid-phase microextraction
SPTD	short-path thermal desorption
SRM	selected reaction monitoring
TD	thermal desorption

TSD   thermoionic specific detection/detector

FIBRES

CAR	carboxen
CW	carbowax
DVB	divinylbenzene
PA	polyacrilate
PDMS	polydimethylsiloxane
TR	template resin

COMPOUNDS

ADBI	celestolide (4-acetyl-1,1-dimethyl-6-tert-butylindane)
AHMI	phantolide (6-acetyl-1,1,2,3,3,5-hexamethyl-indane)
AHTN	tonalide (7-acetyl-1,1,3,4,4,6-hexamethyl-tetraline)
ATII	traseolide, (5-acetyl-1,1,2,6-tetramethyl-3-iso-propyldihydroindane)
BBP	butylbenzyl phthalate
BECDIP	1-bromo-3-ethoxycarbonyloxy-1,2-diiodo-1-propene
BFRs	brominated flame retardants
BTBPE	bis-(2,4,6-tribromophenoxy)ethane
BTEX	mixture of benzene, toluene, ethylbenzene and xylenes
CPIP	1-(4-chlorophenyl)-3-iodopropargylformal
DBP	dibutyl phthalate
DDT	dichloro-diphenyl-trichloroethane
DeBDethane	decabromodiphenyl ethane
DEHP	bis-(2-ethylhexyl) phthalate
DEP	diethyl phthalate
DIBP	diisobutyl phthalate
DIDP	diisodecyl phthalate
DINP	diisononyl phthalate
DMP	dimethyl phthalate

DPMI	Cashmeran (6,7-dihydro-1,1,2,3,3-pentamethyl-4(5H)indanone)
EDCs	endocrine disrupting compounds
EtFOSE	perfluorooctane sulfonamidoethanol
HAPs	hazardous air pollutants
HBCD	hexabromocyclododecane
HCHs	hexachlorocyclohexanes
HHCB	galaxolide (1,3,4,6,7,8-hexahydro-4,6,6,7,8,8-hexamethylcyclopenta-(g)-2-benzopyrane
FTOHs	fluorotelomer alcohols
IPBC	3-iodo-2-propynyl-N-butylcarbamate
MeFOSE	perfluorooctane sulfonamidoethanol
MK	musk ketone (4-tert-butyl-3,5-dinitro-2,6-dimethylacetophenone)
MM	musk moskene (4,6-dinitro-1,1,3,3,5-pentamethylindane)
MTBE	methyl tert-buthyl ether
MX	musk xylene (1-tert-butyl-3,5-dimethyl-2,4,6-trinitrobenzene)
OPPs	organic priority pollutants
OPs	organophosphate esters
PAHs	polycyclic aromatic hydrocarbons
PBBs	polybrominated biphenyls
PBDEs	polybrominated diphenyl ethers
PCBs	polychlorinated biphenyls
PCDDs	polychlorodibenzodioxins
PCDFs	polychlorodibenzofurans
POPs	priority organic pollutants
PVC	polyvynylchloride
PFASs	perfluoralkyl sulfonamides
PFCs	perfluorinated compounds
PFOA	perfluorooctanoate
PFOS	perfluorooctane sulfonate
POPs	priority organic pollutants
SVOCs	semivolatile organic compounds
TBBPA	tetrabromobisphenol-A
TBP	tri-n-butyl phosphate
TBT	tributyl tin
TCEP	tris(2-chloroethyl) phosphate

TCMTB	2-(thiocyanomethylthio)benzothiazole
TCPP	tris(2-chloropropyl) phosphate
TDCT	thermal desorption cold trap
THMs	trihalomethanes
TPeP	tripentyl phosphate
TPhP	triphenyl phosphate
TPP	tripropyl phosphate
TPTC	triphenyltin chloride
VOCs	volatile organic compounds

INDEX

A

acetic acid, 11
acetone, 36, 39, 41, 42, 43, 44, 49, 53, 55, 59, 60, 64, 65, 66, 67, 68, 69, 71, 72, 73, 77
acetonitrile, 4, 22, 38, 68, 69, 72
acetophenone, 13
acid, 11, 36, 52, 67
acidity, 22
acrylate, 16
activated carbon, 9, 16, 28, 57
active site, 16, 29
additives, 34
adsorption, 6, 8, 9, 10, 12, 14, 16, 27, 28, 31, 37, 52, 74, 82
aerosols, 1, 44, 74
air emissions, 12
air pollutants, 58, 102
air quality, 8, 58
alcohol, 39
alcohols, 14, 16, 28, 76, 102
aldehydes, 13, 14
alkenes, 29
aluminium, 35, 52
ambient air, 2, 3, 8, 62, 63, 71
amines, 13, 17, 22
applications, 3, 7, 9, 12, 14, 22, 75
aromatic compounds, 25
aromatics, 14, 29

arthropods, 63
atmospheric pressure, 46, 100
atoms, 21, 46
authors, 13, 27, 45, 55, 70, 71, 73, 75
automation, 15

B

background, 10, 52
benzene, 9, 54, 57, 78, 101
biomonitoring, 8
bisphenol, 69
bleeding, 13, 56
brominated flame retardants, 2, 8, 13, 14, 21, 33, 34, 35, 45, 46, 47, 48, 101
buffer, 8
burning, 8
butadiene, 78, 79
by-products, 71

C

calibration, 6, 7, 22, 26, 30, 31, 32, 66, 70
Canada, 46
cancer, 78
candidates, 76
capillary, 21, 28, 59, 78
carbon, 9, 11, 12, 13, 14, 28, 54, 55, 82, 99
carbon materials, 55
carbon nanotubes, 28, 82, 99
carcinogen, 14
chemical properties, 58

chemical structures, 58
chlordanes, 8, 64, 67, 73, 74
chlorinated hydrocarbons, 29
chromatography, 12, 21, 22, 46, 49, 55, 57, 59, 73, 93, 100
classification, 9
cleaning, 34, 72
CO_2, 19
coatings, 15, 17
combustion, 15
competition, 62
complement, 8
components, 29, 30, 78
composition, 1, 5, 10
compounds, vii, 2, 4, 5, 7, 8, 9, 10, 11, 12, 13, 14, 16, 17, 19, 21, 22, 25, 27, 28, 29, 30, 31, 33, 35, 37, 44, 45, 46, 47, 48, 49, 50, 51, 55, 56, 57, 58, 59, 61, 62, 63, 73, 74, 76, 77, 78, 81, 82, 102
comprehension, 9
concentration, 2, 7, 16, 19, 26, 27, 30, 32, 36, 37, 38, 39, 40, 41, 42, 43, 46, 49, 52, 53, 54, 58, 59, 60, 64, 65, 66, 69, 70, 72, 75, 77, 78, 99
condensation, 44
consumption, vii, 1, 34, 44, 81
contaminant, 15, 44
contamination, 5, 15, 25, 33, 51, 52, 61, 62
control, 25, 62, 74, 75
conversion, 44
correlation, 47, 75
cosmetics, vii, 1, 58, 61
cost effectiveness, 15
crops, 62, 75
crystalline, 15
cycling, 22

D

DBP, 53, 55, 57, 101
decomposition, 13, 28, 31, 35, 70
degradation, vii, 6, 10, 13, 29, 35, 45, 48, 74
degradation process, vii
density, 18, 99
Department of Health and Human Services, 91, 95

deposition, 18, 26, 44, 72
desorption, vii, 3, 4, 9, 10, 12, 13, 15, 18, 19, 20, 42, 43, 52, 55, 59, 64, 67, 73, 78, 81, 82, 100, 103
detection, 6, 18, 19, 21, 45, 46, 50, 52, 55, 59, 72, 73, 99, 100, 101
detection techniques, 21
developed countries, vii, 1
dialysis, 7
dichloroethane, 43
diffusion, 1, 7, 9, 15, 26, 30
dimethylsulfoxide, 38, 40
diseases, 58
displacement, 27
distribution, 31, 75, 81
diversity, 62
DOP, 53
drying, 12, 54, 65, 72

E

elastomers, 57
electron, 21, 100
emerging issues, 33
emission, 2, 21, 46, 63
endocrine, vii, 57, 102
environment, 3, 29, 33, 34, 48, 51, 58, 62, 81
environmental conditions, 7, 28
Environmental Protection Agency, 3, 62, 83, 87, 90, 100
environmental tobacco, 9, 81, 97, 99
EPA, 3, 4, 13, 14, 22, 58, 62, 83, 87, 90, 95, 100
equilibrium, 7, 15, 31, 32, 42, 43, 44, 48
equipment, 1, 2, 19, 34, 48, 51
ESI, 38, 46, 100
ethers, 34, 45
ethyl acetate, 60, 64, 65, 66, 68, 72, 77
European Commission, 57
European Union, 3, 34
evaporation, 20, 65
exposure, vii, 1, 2, 15, 34, 35, 47, 57, 58, 62, 63, 74, 75, 78, 81
extraction, 4, 7, 15, 16, 18, 19, 20, 22, 29, 31, 32, 35, 36, 37, 38, 40, 41, 42, 43, 44,

49, 50, 51, 52, 54, 55, 59, 65, 66, 68, 69, 72, 81, 99, 100

F

film thickness, 21, 59
films, 39, 44
filters, 9, 35, 44, 49, 55, 57, 70, 99
filtration, 38, 40, 49, 64, 66
flame, vii, 2, 31, 33, 34, 35, 36, 44, 46, 48, 49, 50, 81, 100
flame retardants, vii, 2, 31, 33, 34, 35, 36, 44, 46, 48, 49, 50, 81
flammability, 33
fluid, 19
fluid extract, 19
focusing, vii, 20
food, 34, 57, 58
formaldehyde, 90
furniture, 13, 33, 34, 35, 56

G

gas dilution, 30
gases, 1, 15, 30
gel, 10, 13, 44, 52, 59, 64, 70, 100
gel permeation chromatography, 44, 100
Germany, 7
gold, 30
GPC, 36, 38, 44, 100
greenhouse gases, 10
groundwater, 62
groups, 21, 29

H

health, 1, 48, 57, 58, 62, 81
health effects, 1, 58, 81
heating, 15
height, 68, 71
hexabromocyclododecane (HBCD), 33
hexachlorocyclohexanes, 74, 102
hexane, 4, 36, 37, 38, 39, 40, 41, 42, 49, 53, 54, 55, 59, 60, 64, 65, 66, 67, 68, 70, 72
human exposure, 1, 34, 47, 48
humidity, vii, 2, 7, 12, 13, 27, 70
Hunter, 86, 91, 93

hydrocarbons, 97
hydrogen, 61
hygiene, 9

I

identification, 22, 45, 49
impurities, 10, 19, 28, 79
ingestion, 34, 48
insecticide, 31, 68, 71
interface, 18, 99
International Labour Organisation, 95
internet, 33
ionization, 21, 45, 46, 49, 100
ions, 55
irradiation, 20
isomers, 12, 72
isoprene, 78

K

ketones, 4, 14, 28
kindergartens, 1, 57, 61
KOH, 36, 37, 38, 39, 44
Kuwait, 46

L

light conditions, 35
line, 15, 18, 21, 72, 82
linearity, 32
liquid chromatography, 7, 100
Luo, 89

M

magnesium, 57
manipulation, 55
manufacturer, 16
mass spectrometry, 18, 49, 72, 93, 100
matrix, 1, 15, 29, 31, 66, 70, 73
measurement, 26, 78
membranes, 5, 8, 18, 71, 82
Mercury, 88
metabolites, 65, 72, 73, 75
meter, 31, 63
methanol, 13, 19, 37, 38, 54, 57, 65, 72

microenvironments, vii, 1, 46, 81
microwave radiation, 82
microwaves, 7, 19
migration, 51
model, 8, 14, 74, 82
moisture, 6, 15
molecular mass, 48
molecular weight, 17, 22
molecules, 13, 27, 29, 46

N

Na_2SO_4, 12, 54
NaCl, 54
naphthalene, 78
National Bureau of Standards, 30, 100
National Institute for Occupational Safety and Health, 3, 83, 91, 100
nicotine, 78
nitrogen, 21, 22, 29, 30, 32, 49, 52, 97
nitrogen oxides, 29
N-N, 4
nonane, 38, 40
North America, 34
nutrient transfer, 22

O

oil, 76
olefins, 13
operator, 30
optimization, 45
order, 5, 9, 14, 20, 34, 49, 62
organic chemicals, 57, 76
organic compounds, vii, 2, 3, 12, 28, 71, 102, 103
organic peroxides, 19
organic polymers, 9, 12
organic solvents, 52
organotin compounds, 2, 57, 81
oxidation, 62
ozone, 2, 13, 29

P

paints, vii, 1, 34
parameter, 26, 28, 29

parameters, 15, 25, 26, 27, 28, 45
passive, 5, 7, 8, 9, 15, 16, 18, 20, 26, 31, 44, 71, 76, 82, 99
pathways, 2
percentile, 74
performance, vii, 3, 12, 14, 25, 28, 70, 81
permeable membrane, 7
permeation, 7, 18, 32
pesticide, 32, 62, 63, 71, 75
petroleum, 39, 40, 41, 44, 77
phenol, 78
phosphorous, 50
phosphorus, 21, 49
photoirradiation, 29
phthalates, 2, 25, 51, 52, 55, 57, 61, 81
plasma, 16, 18, 72, 75
plastics, 33, 35, 48
polarity, 27
pollutants, vii, 1, 2, 3, 5, 8, 9, 15, 19, 21, 25, 48, 51, 58, 71, 76, 81, 102
pollution, vii, 1, 2, 8, 58
polybrominated biphenyls, 33, 81, 102
polybrominated diphenyl ethers, 2, 33, 81, 93, 102
polycyclic aromatic hydrocarbon, 2, 102
polydimethylsiloxane, 101
polymer, 5, 7, 16, 51
polymeric materials, 51
polymerization, 12
polymers, 13, 51, 57
polypropylene, 18, 72
polyurethane, 7, 13, 14, 16, 35, 44, 47, 48, 49, 52, 55, 57, 59, 70, 76, 99
polyurethane foam, 7, 13, 14, 35, 44, 47, 48, 49, 52, 55, 57, 59, 70, 76, 99
polyvinyl chloride, 57
pressure, 7, 30, 100
production, 12, 33, 34, 35, 51, 58, 62, 81, 95
programming, 22, 56
project, 1
properties, 22, 26, 27, 58, 62
protocol, 25, 34
PTFE, 31, 70, 71
PUFs, 70
pumps, 6, 9, 15

Index

purification, 19, 64
PVC, 51, 57, 102
pyrolysis, 9

Q

quality assurance, 34, 99
quality control, 99
quartz, 9, 31, 48, 55, 57, 70
quaternary ammonium, 73

R

range, 6, 12, 14, 16, 23, 26, 30, 32, 35, 45, 46, 50, 51, 56, 57, 62, 70, 71, 72, 82
reactions, 29
reactive sites, 28
reagents, 51
reason, 3, 13, 35
recommendations, iv, 35
recovery, 14, 22, 25, 27, 28, 29, 32, 59, 61, 66, 70
regulations, 63
regulatory requirements, 2
relationship, 1
reliability, 62
residues, 71, 74
resins, 11, 13, 14, 70, 71
restaurants, 78, 79
retention, 10, 12, 26, 28, 29, 70, 73
retention volume, 26
rights, iv
rings, 52
risk, 61, 72, 78
robustness, 44, 45
Royal Society, 89
rubber, 10, 32

S

sampling, 2, 4, 5, 6, 7, 8, 9, 12, 13, 14, 15, 16, 18, 20, 22, 25, 26, 27, 28, 29, 30, 31, 35, 44, 46, 48, 49, 50, 51, 55, 56, 57, 58, 61, 67, 70, 71, 73, 75, 76, 78, 82, 92
school, vii, 1, 51, 81
selecting, 16
selectivity, 16, 21, 45, 49, 50, 73

semi-permeable membrane, 7
sensitivity, 5, 15, 16, 18, 21, 22, 25, 26, 32, 44, 45, 49, 52, 55, 73, 82
septum, 32, 67
SII, 11
silica, 10, 13, 15, 36, 37, 38, 39, 40, 41, 44, 48, 49, 52, 59, 60, 64, 70
silicones, 55
smoke, 9, 81, 97, 99
sodium, 29, 52, 53, 65, 72
soil, 13, 62
solid phase, 22, 100
solvents, 35, 52, 55, 71, 72
sorption, 9, 12, 13
Spain, 61
species, 9, 44
stability, 10, 12, 25, 26, 27, 28, 70
stabilizers, 57
standard deviation, 30, 32, 78
standardization, 3, 31
standards, 3, 15, 30, 33, 44, 45, 50, 61
steel, 4, 5, 8, 18
storage, 6, 12, 25, 27, 28, 70
styrene, 35, 78
substitutes, 34, 63
substitution, 55
sulfonamides, 76, 102
surface area, 14, 82
surface treatment, 76
surrogates, 36, 37, 38, 39, 40, 41, 60, 69, 77
Sweden, 8, 46, 93

T

TBP, 41, 42, 43, 48, 51, 102
teachers, 1
teaching, 46
temperature, vii, 2, 7, 9, 13, 16, 22, 26, 27, 28, 45, 48, 75, 100
terpenes, 14, 29
textiles, 33, 34
thermal degradation, 45
thermal stability, 10
thermal treatment, 34, 52
time periods, 44
tin, 57, 102

tobacco, 15, 78
tobacco smoke, 78
toluene, 4, 9, 18, 36, 38, 39, 53, 55, 66, 67, 72, 78, 101
toxic effect, 62, 76, 81
toxicity, 63
transformation, 74
transformation product, 74
transport, 6, 61
tributyl phosphate, 50
tricresyl phosphate, 12

U

ultrasound, 66
United Kingdom, 8
United Nations, 95
United States, 25, 100
upholstery, 76
UV light, 35

UV-radiation, 62

V

vacuum, 31
vegetables, 71, 75
velocity, 7
ventilation, vii, 2, 75
volatility, 6, 13, 14, 27, 51, 82
volatilization, 34
vulnerable people, 1

W

wool, 31, 40, 41, 49, 64, 65, 66, 72
work environment, 46
workers, 15, 71, 74, 75
workplace, 3, 7, 9, 12, 61, 63, 70, 71